DNA MICROARRAYS AND GENE EXPRESSION

From experiments to data analysis and modeling

Massive data acquisition technologies, such as genome sequencing, high-throughput drug screening, and DNA arrays are in the process of revolutionizing biology and medicine. Using the mRNA of a given cell, at a given time, under a given set of conditions, DNA microarrays can provide a snapshot of the level of expression of all the genes in the cell. Such snapshots can be used to study fundamental biological phenomena such as development or evolution, to determine the function of new genes, to infer the role that individual genes or group of genes may play in diseases, and to monitor the effect of drugs and other compounds on gene expression. This interdisciplinary introduction to DNA arrays will be essential reading for researchers wanting to take advantage of this powerful new technology.

PIERRE BALDI is Professor and Director of the Institute for Genomics and Bioinformatics in the Department of Information and Computer Science and in the Department of Biological Chemistry at the University of California, Irvine.

WES HATFIELD is a Professor in the Department of Microbiology and Molecular Genetics in the College of Medicine and the Department of Chemical Engineering and Material Sciences in the School of Engineering at the University of California, Irvine.

DNA MICROARRAYS AND GENE EXPRESSION

From experiments to data analysis and modeling

PIERRE BALDI

University of California, Irvine

and

G. WESLEY HATFIELD

University of California, Irvine

CAMBRIDGE
UNIVERSITY PRESS

PUBLISHED BY THE PRESS SYNDICATE OF THE UNIVERSITY OF CAMBRIDGE
The Pitt Building, Trumpington Street, Cambridge, United Kingdom

CAMBRIDGE UNIVERSITY PRESS
The Edinburgh Building, Cambridge CB2 2RU, UK
40 West 20th Street, New York, NY 10011-4211, USA
477 Williamstown Road, Port Melbourne, VIC 3207, Australia
Ruiz de Alarcón 13, 28014 Madrid, Spain
Dock House, The Waterfront, Cape Town 8001, South Africa

http://www.cambridge.org

First published 2002

Printed in the United Kingdom at the University Press, Cambridge

Typeface Times NR MT 10/13pt   System QuarkXPress™   [SE]

A catalogue record for this book is available from the British Library

Library of Congress Cataloguing in Publication data
Baldi, Pierre.
DNA microarrays and gene expression / Pierre Baldi and G. Wesley Hatfield.
p. cm.
Includes bibliographical references and index.
ISBN 0 521 80022 6
1. DNA microarrays. 2. Gene expression. I. Hatfield, G. Wesley, 1940– II. Title.
QP624.5.D726 .B353 2002
572.8′65 – dc21   2001052862

ISBN 0 521 80022 6 hardback

Contents

Preface

A number of array-based technologies have been developed over the last several years, and technological development in this area is likely to continue at a brisk pace. These technologies include DNA, protein, and combinatorial chemistry arrays. So far, DNA arrays designed to determine gene expression levels in living cells have received the most attention. Since DNA arrays allow simultaneous measurements of thousands of interactions between mRNA-derived target molecules and genome-derived probes, they are rapidly producing enormous amounts of raw data never before encountered by biologists. The bioinformatics solutions to problems associated with the analysis of data on this scale are a major current challenge.

Like the invention of the microscope a few centuries ago, DNA arrays hold promise of transforming biomedical sciences by providing new vistas of complex biological systems. At the most basic level, DNA arrays provide a snapshot of all of the genes expressed in a cell at a given time. Therefore, since gene expression is the fundamental link between genotype and phenotype, DNA arrays are bound to play a major role in our understanding of biological processes and systems ranging from gene regulation, to development, to evolution, and to disease from simple to complex. For instance, DNA arrays should play a role in helping us to understand such difficult problems as how each of us develops from a single cell into a gigantic supercomputer of roughly 10^{15} cells, and why some cells proliferate in an uncontrolled manner to cause cancer.

One notable difference between modern DNA array technology and the seventeenth-century microscope, however, is in the output produced by these technologies. In both cases, it is an image. But unlike the image one sees through a microscope, an array image is not interpretable by the human eye. Instead, each individual feature of the DNA array image must be measured and stored in a large spreadsheet of numbers with tens to tens-of-thousands of rows associated with gene probes, and as many columns or experimental conditions as the experimenter is willing to collect. As a side note, this may change in the future and one could envision simple diagnostic arrays that can be read directly by a physician.

Clearly, the scale and tools of biological research are changing. The storage, retrieval, interpretation, and integration of large volumes of data generated by DNA arrays and other high-throughput technologies, such as genome sequencing and mass spectrometry, demand increasing reliance on computers and evolving computational methods. In turn, these demands are effecting fundamental changes in how research is done in the life sciences and the culture of the biological research community. It is becoming increasingly important for individuals from both life and computational sciences to work together as integrated research teams and to train future scientists with interdisciplinary skills. It is inevitable that as we enter farther into the genomics era, single-investigator research projects typical of research funding programs in the biological sciences will become less prevalent, giving way to more interdisciplinary approaches to complex biological questions conducted by multiple investigators in complementary fields. Statistical methods, in particular, are essential for the interpretation of high-throughput genomic data. Statistics is no longer a poor province of mathematics. It is rapidly becoming recognized as the central language of sciences that deal with large amounts of data, and rely on inferences in an uncertain environment.

As genomic technologies and sequencing projects continue to advance, more and more emphasis is being placed on data analysis. For example, the identification of the function of a gene or protein depends on many things including structure, expression levels, cellular localization, and functional neighbors in a biochemical pathway that are often co-regulated and/or found in neighboring regions along the chromosome. Clearly then, establishing the function of new genes can no longer depend on sequence analysis alone but requires taking into account additional sources of information including phylogeny, environment, molecular and genomic structure, and metabolic and regulatory networks. By contributing to the understanding of these networks, DNA arrays already are playing a significant role in the annotation of gene function, a fundamental task of the genomics era. At the same time, array data must be integrated with sequence data, with structure and function data, with pathway data, with phenotypic and clinical data, and so forth. New biological discoveries will depend strongly on our ability to combine and correlate these diverse data sets along multiple dimensions and scales. Basic research in bioinformatics must deal with these issues of systems and integrative biology in a situation where the amount of data is growing exponentially.

As these challenges are met, and as DNA array technologies progress, incredible new insights will surely follow. One of the most striking results of the Human Genome Project is that humans probably have only on the order of twice the number of genes of other metazoan organisms such as the fly. While these numbers are still being revised, it is clear that biological complexity does not come from sheer gene number but from other sources. For instance, the number of gene products and their interactions can be greatly amplified by mechanisms such as alternative mRNA splicing, RNA editing, and post-translational protein modifications. On top of this, additional levels of complexity are generated by genetic and biochemical networks responsible for the integration of multiple biological processes as well as the effects of the environment on living cells. Surely, DNA and protein array technologies will contribute to the unraveling of these complex interactions. At a time when human cloning and organ regeneration from stem cells are on the horizon, arrays should help us to further understand the old but still largely unanswered questions of nature versus nurture and perhaps strike a new balance between the reductionist determinism of molecular biology and the role of chance, epigenetic regulation, and environment on living systems. However, while arrays and other high-throughput technologies will provide the data, new bioinformatics innovations must provide the methods for the elucidation of these complex interactions.

As we progress into the genomics era, it is anticipated that DNA array technologies will assume an increasing role in the investigation of evolution. For example, DNA array studies could shed light on mechanisms of evolution directly by the study of mRNA levels in organisms that have fast generation times and indirectly by giving us a better understanding of regulatory circuits and their structure, especially developmental regulatory circuits. These studies are particularly important for understanding evolution for two obvious reasons: first, genetic adaptation is very constrained since most non-neutral mutations are disadvantageous; and second, simple genetic changes can serve as "amplifiers" in the sense that they can produce large developmental changes, for instance doubling the number of wings in a fly.

On the medical side, DNA arrays ought to help us better understand complex issues concerning human health and disease. Among other things, they should help us tease out the effects of environment and life style, including drugs and nutrition, and help usher in the individualized molecular medicine of the future. For example, daily doses of vitamin C recommended in the literature vary over three orders of magnitude. In fact, the optimal dose for all nutritional supplements is unknown. Information from DNA array studies should help define and quantify the impact of these supplements on human health. Furthermore, although we are accustomed to the expression "the human body", large response variabilities among individuals due to genetic and environmental differences are observed. In time, the information obtained from DNA array studies should help us tailor nutritional intake and therapeutic drug doses to the makeup of each individual.

Throughout the second half of the twentieth century, molecular biologists have predominantly concentrated on single-gene/single-protein studies. Indeed, this obsessive focus on working with only one variable at a time while suppressing all others in *in vitro* systems has been a hallmark of molecular biology and the foundation for much of its success. As we enter into the genomics era this basic paradigm is shifting from the study of single-variable systems to the study of complex interactions. During this same period, cell biologists have been following mRNA and/or protein levels during development, and much of what we know about development has been gathered with techniques like *in situ* hybridization that have allowed us to define gene regulatory mechanisms and to follow the expression of individual genes in multiple tissues. DNA arrays give us the additional ability to follow the expression levels of all of the genes in the cells of a given tissue at a given time.

As old and new technologies join forces, and as computational scientists and biologists embrace the high-throughput technologies of the genomics era, the trend will be increasingly towards a systems biology approach that simultaneously studies tens of thousands of genes in multiple tissues under a myriad of experimental conditions. The goal of this systems biology approach is to understand systems of ever-increasing complexity ranging from intracellular gene and protein networks, to tissue and organ systems, to the dynamics of interactions between individuals, populations, and their environments. This large-scale, high-throughput, interdisciplinary approach enabled by genomic technologies is rapidly becoming a driving force of biomedical research particularly apparent in the biotechnology and pharmaceutical industries. However, while the DNA array will be an important workhorse for the attainment of these goals, it should be emphasized that DNA array technology is still at an early stage of development. It is cluttered with heterogeneous technologies and data formats as well as basic issues of noise, fidelity, calibration, and statistical significance that are still being sorted out. Until these issues are resolved

and standardized, it will not be possible to define the complete genetic regulatory network of even a well-studied prokaryotic cell. In the meantime,
most progress will continue to come from focused incremental studies that
look at specific networks and specific interacting sets of genes and proteins
in simple model organisms, such as the bacterium *Escherichia coli* or the
yeast *Saccharomyces cerevisiae*.

In short, the promise of DNA arrays is to help us untangle the extremely
complex web of relationships among genotypes, phenotypes development,
environment, and evolution. On the medical side, DNA arrays ought to
help us understand disease, create new diagnostic tools, and help usher in
the individualized molecular medicine of the future. DNA array technology is here and progressing at a rapid pace. The bioinformatics methods to
process, analyze, interpret, and integrate the enormous volumes of data to
be generated by this technology are coming.

Audience and prerequisites

In 1996, Hillary Rodham Clinton published a book titled *It Takes a Village*.
This book discusses the joint responsibilities of different segments of a
community for raising a child. Just like it takes individuals with different
talents to raise a child "it takes a village" to do genomics. More precisely, it
takes an ongoing dialog and a two-way flow of information and ideas
between biologists and computational scientists to develop the designs and
methods of genomic experiments that render them amenable to rigorous
analysis and interpretation. This book seeks to foster these interdisciplinary
interactions by providing in-depth descriptions of DNA microarray technologies that will provide the information necessary for the design and execution of DNA microarray experiments that address biological questions of
specific interest. At the same time, it provides the details and discussions of
the computational methods appropriate for the analysis of DNA microarray data. In this way, it is anticipated that the computational scientists will
benefit from learning experimental details and methods, and that the biologist will benefit from the discussions of the methods for the analysis and
interpretation of data that results from these high dimensional experiments.

At this time, most biologists depend on their computational colleagues
for the development of data analysis methods and the computational scientists depend upon their biologist colleagues to perform experiments that
address important biological questions and to generate data. Since this
book is directed to both fields, we hope that it will serve as a catalyst to facilitate these critical interactions among researchers of differing talents and

expertise. In other words, we have tried to write a book for all members of the genomic village. In this effort, we anticipate that a biologist's full appreciation of the more computationally intense sections might require consultation with a computational colleague, and that the computational scientists will benefit from discussions concerning experimental details and strategies with the biologist. For the most part, however, this book should be generally comprehensible by either a biologist or a computational scientist with a basic background in biology and mathematics. It is written for students, mostly at the graduate but possibly at the undergraduate level, as well as academic and industry researchers with a diverse background along a broad spectrum from computational to biomedical sciences. It is perhaps fair to say that the primary readers we have in mind are researchers who wish to carry out and interpret DNA array experiments. To this end, we have endeavored to provide succinct explanations of core concepts and techniques.

Content and general outline of the book

We have tried to write a comprehensive but reasonably concise introductory book that is self-contained and gives the reader a good sense of what is available and feasible today. We have not attempted to provide detailed information about all aspects of arrays. For example, we do not describe how to build your own array since this information can be obtained from many other sources, ranging from Patrick Brown's web site at Stanford University to a book by Schena *et al.*[1] Instead, we focus on DNA array experiments, how to plan and execute them, how to analyze the results, what they are good for, and pitfalls the researcher may encounter along the way.

The topics of this book reflect our personal biases and experiences. A significant portion of the book is built on material from articles we have written, our unpublished observations, and talks and tutorials we have presented at several conferences and workshops. While we have tried to quote relevant literature, we have concentrated our main effort on presenting the basic concepts and techniques and illustrating them with examples. The main focus of this book is on methods – how to design, execute and interpret a gene expression profiling experiment in a way that remains flexible and open to future developments.

In Chapter 1 we present a brief history of genomics that traces some of

---

[1] Brown, P. http://cmgm.stanford.edu/pbrown; Schena, M. (ed.) *Microarray Biochip Technology*. 2000. Eaton Publishing Co., Natick, MA.

the milestones over the past 50 years or so that have ushered us into the genomics era. This history emphasizes the technological breakthroughs – and the authors' bias towards the importance of the model organism *Escherichia coli* in the development of the paradigms of modern molecular biology – that have led us from the enzyme period to the genomics era.

In Chapter 2 we describe the various DNA array technologies that are available today. These technologies range from *in situ* synthesized arrays such as the Affymetrix GeneChip™, to pre-synthesized nylon membrane and glass slide arrays, to newer technologies such as electronic and bead-based arrays.

In Chapter 3 we describe the methods, technology, and instrumentation required for the acquisition of data from DNA arrays hybridized with radioactive-labeled or fluorescent-labeled targets.

In Chapter 4 we consider issues important for the design and execution of a DNA array experiment with special emphasis on problems and pitfalls encountered in gene expression profiling experiments. Special consideration is given to experimental strategies to deal with these problems and methods to reduce experimental and biological sources of variance.

In Chapter 5 we deal with the first level of statistical analysis of DNA array data for the identification of differentially expressed genes. Due to the large number of measurements from a single experiment, high levels of noise, and experimental and biological variabilities, array data is best modeled and analyzed using a probabilistic framework. Here we review several approaches and develop a practical Bayesian statistical framework to effectively address these problems to infer gene changes. This framework is applied to experimental examples in Chapter 7.

In Chapter 6 we move to the next level of statistical analysis involving the application of visualization, dimensionality reduction, and clustering methods to DNA array data. The most popular dimensionality and clustering methods and their advantages and disadvantages are surveyed. We also examine methods to leverage array data to identify DNA genomic sequences important for gene regulation and function. Mathematical details for Chapters 5 and 6 are presented in Appendix B.

In Chapter 7 we present a brief survey of current DNA array applications and lead the reader through a gene expression profiling experiment taken from our own work using pre-synthesized (nylon membrane) and *in situ* synthesized (Affymetrix GeneChip™) DNA arrays. Here we describe the use of software tools that apply the statistical methods described in Chapters 5 and 6 to analyze and interpret DNA array data. Special emphasis is given to methods to determine the magnitude and sources of experimental errors and how to use this information to determine global false positive rates and confidence levels.

Chapter 8 covers several aspects of what is coming to be known as systems biology. It provides an overview of regulatory, metabolic, and signaling networks, and the mathematical and software tools that can be used for their investigation with an emphasis on the inference and modeling of gene regulatory networks.

The appendices include explicit technical information regarding: (A) protocols, such as RNA preparation and target labeling methods, for DNA array experiments; (B) additional mathematical details about, for instance, support vector machines; (C) a section with a brief overview of current database resources and other information that are publicly available over the Internet, together with a list of useful web sites; and (D) an introduction to CyberT, an online program for the statistical analysis of DNA array data.

Finally, a word on terminology. Throughout the book we have used for the most part the word "array" instead of "microarray" for two basic reasons: first, in our minds DNA arrays encompass DNA microarrays; second, at what feature density or physical size an array becomes a microarray is not clear. Also, the terms "probe" and "target" have appeared interchangeably in the literature. Here we keep to the nomenclature for probes and targets of northern blots familiar to molecular biologists; we refer to the nucleic acid attached to the array substrate as the "probe" and the free nucleic acid as the "target".

Acknowledgements

Many colleagues have provided us with input, help, and support. At the risk of omitting many of them, we would like to thank in particular Suzanne B. Sandmeyer who has been instrumental in developing a comprehensive program in genomics and the DNA Array Core Facility at UCI, and who has contributed in many ways to this work. We would like to acknowledge our many colleagues from the UCI Functional Genomics Group for the tutelage they have given us at their Tuesday morning meetings, in particular: Stuart Arfin, Lee Bardwell, J. David Fruman, Steven Hampson, Denis Heck, Dennis Kibler, Richard Lathrop, Anthony Long, Harry Mangalam, Calvin McLaughlin, Ming Tan, Leslie Thompson, Mark Vawter, and Sara Winokur. Outside of UCI, we would like to acknowledge our collaborators Craig J. Benham, Rob Gunsalus, and David Low. We would like also to thank Wolfgang Banzhaf, Hamid Bolouri, Hidde de Jong, Eric Mjolsness,

James Nowick, and Padhraic Smyth for helpful discussions. Hamid and Hidde also provided graphical materials, James and Padhraic provided feedback on an early version of Chapter 8, and Wolfgang helped proofread the final version. Additional material was kindly provided by John Weinstein and colleagues, and by Affymetrix including the image used for the cover of the book. We would like to thank our excellent graduate students Lorenzo Tolleri, Pierre-François Baisnée, and Gianluca Pollastri, and especially She-pin Hung, for their help and important contributions. We gratefully acknowledge support from Sun Microsystems, the Howard Hughes Medical Institute, the UCI Chao Comprehensive Cancer Center, the UCI Institute for Genomics and Bioinformatics (IGB), the National Science Foundation, the National Institutes of Health, a Laurel Wilkening Faculty Innovation Award, and the UCI campus administration, as well as a GAANN (Graduate Assistantships in Areas of National Need Program) and a UCI BREP (Biotechnology Research and Education Program) training grant in Functional and Computational Genomics. Ann Marie Walker at the IGB helped us with the last stages of this project. We would like also to thank our editors, Katrina Halliday and David Tranah at Cambridge University Press, especially for their patience and encouragement, and all the staff at CUP who have provided outstanding editorial help. And last but not least, we wish to acknowledge the support of our friends and families.

1

A brief history of genomics

From time to time new scientific breakthroughs and technologies arise that forever change scientific practice. During the last 50 years, several advances stand out in our minds that – coupled with advances in the computational and computer sciences – have made genomic studies possible. In the brief history of genomics presented here we review the circumstances and consequences of these relatively recent technological revolutions.

Our brief history begins during the years immediately following World War II. It can be argued that the enzyme period that preceded the modern era of molecular biology was ushered in at this time by a small group of physicists and chemists, R. B. Roberts, P. H. Abelson, D. B. Cowie, E. T. Bolton, and J. R. Britton in the Department of Terrestrial Magnetism of the Carnegie Institution of Washington. These scientists pioneered the use of radioisotopes for the elucidation of metabolic pathways. This work resulted in a monograph titled *Studies of Biosynthesis in* Escherichia coli that guided research in biochemistry for the next 20 years and, together with early genetic and physiological studies, helped establish the bacterium *E. coli* as a model organism for biological research [1]. During this time, most of the metabolic pathways required for the biosynthesis of intermediary metabolites were deciphered and biochemical and genetic methods were developed to identify and characterize the enzymes involved in these pathways.

Much in the way that genomic DNA sequences are paving the way for the elucidation of global mechanisms for genetic regulation today, the biochemical studies initiated in the 1950s that were based on our technical abilities to create isotopes and radiolabel biological molecules paved the way for the discovery of the basic mechanisms involved in the regulation of metabolic pathways. Indeed, these studies defined the biosynthetic pathways for the building blocks of macromolecules such as proteins and

1

nucleic acids and led to the discovery of mechanisms important for metabolic regulation such as end product inhibition, allostery, and modulation of enzyme activity by protein modifications. However, major advances concerning the biosynthesis of macromolecules awaited another breakthrough, the description of the structure of the DNA helix by James D. Watson and Francis H. C. Crick in 1953 [2]. With this information, the basic mechanisms of DNA replication, protein synthesis, gene expression, and the exchange and recombination of genetic material were rapidly unraveled.

During the enzyme period, geneticists around the world were using the information provided by biochemists to develop model systems such as bacteria, fruit flies, yeast, and mice for genetic studies. In addition to establishment of the basic mechanisms for protein-mediated regulation of gene expression by F. Jacob and J. Monod in 1961 [3], these genetic studies led to fundamental discoveries that were to spawn yet another major change in the history of molecular biology. This advance was based on studies designed to determine why *E. coli* cells once infected by a bacteriophage were immune to subsequent infection. These seemingly esoteric investigations led by Daniel Nathans and Hamilton Smith [4] resulted in the discovery of new types of enzymes, restriction endonucleases and DNA ligases, capable of cutting and rejoining DNA at sequence-specific sites. It was quickly recognized that these enzymes could be used to construct recombinant DNA molecules composed of DNA sequences from different organisms. As early as 1972 Paul Berg and his colleagues at Stanford University developed an animal virus, SV40, vector containing bacteriophage lambda genes for the insertion of foreign DNA into *E. coli* cells [5]. Methods of cloning and expressing foreign genes in *E. coli* have continued to progress until today they are fundamental techniques upon which genomic studies and the entire biotechnology industry are based.

The recent history of genomics also has been driven by technological advances. Foremost among these advances were the methodologies of the polymerase chain reaction (PCR) and automated DNA sequencing. PCR methods allowed the amplification of usable amounts of DNA from very small amounts of starting material. Automated DNA sequencing methods have progressed to the point that today the entire DNA sequence of microbial genomes containing several million base pairs can be obtained in less than one week. These accomplishments set the stage for the human genome project.

As early as 1984 the small genomes of several microbes and bacteriophages had been mapped and partially sequenced; however, the modern era

of genomics was not formally initiated until 1986 at an international conference in Santa Fe, New Mexico sponsored by the Office of Health and
Environmental Research[1] of the US Department of Energy. At this
meeting, the desirability and feasibility of implementing a human genome
program was unanimously endorsed by leading scientists from around the
world. This meeting led to a 1988 study by the National Research Council
titled *Mapping and Sequencing the Human Genome* that recommended the
United States support a human genome program and presented an outline
for a multiphase plan. In that same year, three genome research centers
were established at the Lawrence Berkeley, Lawrence Livermore, and Los
Alamos national laboratories. At the same time, under the leadership of
Director James Wyngaarden, the National Institutes of Health established
the Office of Genome Research which in 1989 became the National Center
for Human Genome Research, directed by James D. Watson. The next ten
years witnessed rapid progress and technology developments in automated
sequencing methods. These technologies led to the establishment of large-
scale DNA sequencing projects at many public research institutions around
the world such as the Whitehead Institute in Boston, MA and the Sanger
Centre in Cambridge, UK. These activities were accompanied by the rapid
development of computational and informational methods to meet challenges created by an increasing flow of data from large-scale genome
sequencing projects.

In 1991 Craig Venter at the National Institutes of Health developed a
way of finding human genes that did not require sequencing of the entire
human genome. He relied on the estimate that only about 3 percent of the
genome is composed of genes that express messenger RNA. Venter suggested that the most efficient way to find genes would be to use the processing machinery of the cell. At any given time, only part of a cell's DNA is
transcriptionally active. These "expressed" segments of DNA are converted and edited by enzymes into mRNA molecules. Using an enzyme,
reverse transcriptase, cellular mRNA fragments can be transcribed into
complementary DNA (cDNA). These stable cDNA fragments are called
expressed sequence tags, or ESTs. Computer programs that match overlapping ends of ESTs were used to assemble these cDNA sequences into longer
sequences representing large parts, or all, of many human genes. In 1992,
Venter left NIH to establish The Institute for Genomic Research, TIGR. By
1995 researchers in public and private institutions had isolated over 170 000

---

[1] Changed in 1998 to the Office of Biological and Environmental Research of the
Department of Energy.

ESTs, which were used to identify more than half of the then estimated 60 000 to 80 000 genes in the human genome.[2] In 1998, Venter joined with Perkin-Elmer Instruments (Boston, MA) to form Celera Genomics (Rockville, MD).

With the end in sight, in 1998 the Human Genome Program announced a plan to complete the human genome sequence by 2003, the 50th anniversary of Watson and Crick's description of the structure of DNA. The goals of this plan were to:

- Achieve coverage of at least 90% of the genome in a working draft based on mapped clones by the end of 2001.
- Finish one-third of the human DNA sequence by the end of 2001.
- Finish the complete human genome sequence by the end of 2003.
- Make the sequence totally and freely accessible.

On June 26, 2000, President Clinton met with Francis Collins, the Director of the Human Genome Program, and Craig Venter of Celera Genomics to announce that they had both completed "working drafts" of the human genome, nearly two years ahead of schedule. These drafts were published in special issues of the journals *Science* and *Nature* early in 2001 [6, 7] and the sequence is online at the National Center for Biotechnology Information (NCBI) of the Library of Medicine at the National Institutes of Health

As of this writing, the NCBI databases also contain complete or in progress genomic sequences for ten *Archaea* and 151 bacteria as well as the genomic sequences of eight eukaryotes including: the parasites *Leishmania major* and *Plasmodium falciparum*; the worm *Caenorhabditis elegans*; the yeast *Saccharomyces cerevisiae*; the fruit fly *Drosophila melanogaster*; the mouse *Mus musculus*; and the plant *Arabidopsis thaliana*. Many more genome sequencing projects are under way in private and public research laboratories that are not yet available on public databases. It is anticipated that the acquisition of new genome sequence data will continue to accelerate. This exponential increase in DNA sequence data has fuelled a drive to develop technologies and computational methods to use this information to study biological problems at levels of complexity never before possible.

---

[2] At the present time (September 2001) the estimate of the number of human genes has decreased nearly twofold.

REFERENCES

1. Roberts, R. B., Abelson, P. H., Cowie, D. B., Bolton, E. B., and Britten, J. R. *Studies of Biosynthesis in* Escherichia coli. 1955. Carnegie Institution of Washington, Washington, DC.
2. Watson, J. D., and Crick, F. H. C. A structure for deoxyribose nucleic acid. 1953. *Nature* 171:173.
3. Jacob, F., and Monod, J. Genetic regulatory mechanisms in the synthesis of proteins. 1961. *Journal of Molecular Biology* 3:318–356.
4. Nathans, D., and Smith, H. O. A suggested nomenclature for bacterial host modification and restriction systems and their enzymes. 1973. *Journal of Molecular Biology* 81:419–423.
5. Jackson, D. A., Symons, R. H., and Berg, P. Biochemical method for inserting new genetic information into DNA of simian virus 40: circular SV40 DNA molecules containing lambda phage genes and the galactose operon of *Escherichia coli*. 1972. *Procedings of the National Academy of Sciences of the USA* 69:2904–2909.
6. *Science* Human Genome Issue. 2001. 16 February, vol. 291.
7. *Nature* Human Genome Issue. 2001. 15 February, vol. 409.

2

DNA array formats

Array technologies monitor the combinatorial interaction of a set of molecules, such as DNA fragments and proteins, with a predetermined library of molecular probes. The currently most advanced of these technologies is the use of DNA arrays, also called DNA chips, for simultaneously measuring the level of the mRNA gene products of a living cell. This method, gene expression profiling, is the major topic of this book.

In its most simple sense, a DNA array is defined as an orderly arrangement of tens to hundreds of thousands of unique DNA molecules (probes) of known sequence. There are two basic sources for the DNA probes on an array. Either each unique probe is individually synthesized on a rigid surface (usually glass), or pre-synthesized probes (oligonucleotides or PCR products) are attached to the array platform (usually glass or nylon membranes). The various types of DNA arrays currently available for gene expression profiling, as well as some developing technologies, are summarized here.

In situ synthesized oligonucleotide arrays

The first *in situ* probe synthesis method for manufacturing DNA arrays was the photolithographic method developed by Fodor *et al.* [1] and commercialized by Affymetrix Inc. (Santa Clara, CA). First, a set of oligonucleotide DNA probes (each 25 or so nucleotides in length) is defined based on its ability to hybridize to complementary sequences in target genomic loci or genes of interest. With this information, computer algorithms are used to design photolithographic masks for use in manufacturing the probe arrays. Selected addresses on a photo-protected glass surface are illuminated through holes in the photolithographic mask, the glass surface is flooded with the first nucleotide of the probes to be synthesized at the selected addresses, and photo-chemical coupling occurs at these sites. For example, the addresses on the glass surface for all probes beginning with guanosine are photo-activated and chemically coupled to guanine bases. This step is repeated three more times with masks for all addresses with probes beginning with adenosine, thymine, or cytosine. The cycle is repeated with masks designed for adding the appropriate second nucleotide of each probe. During the second cycle, modified phosphoramidite moieties on each of the nucleosides attached to the glass surface in the first step are light-activated through appropriate masks for the addition of the second base to each growing oligonucleotide probe. This process is continued until unique probe oligonucleotides of a defined length and sequence have been synthesized at each of thousands of addresses on the glass surface (Figure 2.1).

Several companies such as Protogene (Menlo Park, CA) and Agilent Technologies (Palo Alto, CA) in collaboration with Rosetta Inpharmatics (Kirkland, WA) of Merck & Co. Inc. (Whitehouse Station, NJ) have developed *in situ* DNA array platforms through proprietary modifications of a standard piezoelectric (ink-jet) printing process that unlike the manufacturing process for Affymetrix GeneChips™, does not require photolithography. These *in situ* synthesized oligonucleotide arrays are fabricated directly on a glass support on which oligonucleotides up to 60 nucleotides are synthesized using standard phosphoramidite chemistry. The ink-jet printing technology is capable of depositing very small volumes – picoliters per spot – of DNA solutions very rapidly and very accurately. It also delivers spot shape uniformity that is superior to other deposition methods.

Researchers in the Nano-fabrication Center at the University of Wisconsin have developed yet another method for the manufacture of *in situ* synthesized DNA arrays that also does not require photolithographic masks [2]. This technology known as MAS for maskless array synthesizer capitalizes on existing electronic chips used in overhead projection known as digital light processors (DLPs). A DLP is an array of up to 500 000 tiny aluminum mirrors arranged on a computer chip. By electronic manipulation of the mirrors, light can be directed to specific addresses on the surface of a DNA array substrate, thus eliminating the need for expensive photolithographic masks. This technology is being implemented by NimbleGen Systems, LLC (Madison, WI). DNA arrays containing over 307 000 discrete features are currently being synthesized and plans are under way to synthesize a second-generation MAS array containing over 2 million discrete features. The Wisconsin researchers claim that this method will greatly reduce the time and cost for the manufacture of high-density *in situ*

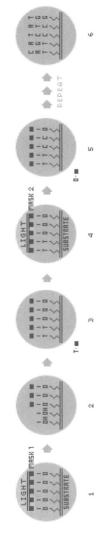

Figure 2.1. The Affymetrix method for the manufacture of *in situ* synthesized DNA microarrays (courtesy of Affymetrix). (1) A photo-protected glass substrate is selectively illuminated by light passing through a photolithographic mask. (2) Deprotected areas are activated. (3) The surface is flooded with a nucleoside solution and chemical coupling occurs at photo-activated positions. (4) A new photolithographic mask pattern is applied. (5) The coupling step is repeated. (6) This process is repeated until the desired set of probes is obtained.

DNA microarray formats

Table 2.1. *Commercial sources for DNA arrays*

Company	Nylon filters	Glass slides	Plastic slides	Chips	Web site
Affymetrix[1,2,3,4,5,6,7,8,18]				X	www.affymetrix.com
Agilent Technologies[18]		X			www.chem.agilent.com
AlphaGene[1,18]		X			www.alphagene.com
Clontech[1,2,3,18]	X	X	X		www.clontech.com
Corning[6]		X			www.corning.com/cmt
Eurogentec[5,6,9,11,12,14,15,16,18]	X	X			www.eurogentec.be
Genomic Solutions[1,2,3]		X			www.genomicsolutions.com
Genotech[1,2]	X				www.genotech.com
Incyte Pharmaceuticals[1,2,3,4,9,10,18]	X	X			www.incyte.com
Invitrogen[1,2,3,6]	X			X	www.invitrogen.com
Iris BioTechnologies[1]					www.irisbiotech.com
Mergen Ltd[1,2,3]		X			www.mergen-ltd.com
Motorola Life Science[1,3,18]		X			www.motorola.com/lifesciences
MWG Biotech[3,6,8,18]		X			www.mwg-biotech.com
Nanogen				X	www.nanogen.com
NEN Life Science Products[1]		X		X	www.nenlifesci.com
Operon Technologies Inc.[1,6,18]		X			www.operon.com
Protogene Laboratories[18]					www.protogene.com
Radius Biosciences[18]		X			www.ultranet.com/~radius
Research Genetics[1,2,3,6]	X				www.resgen.com
Rosetta Inpharmatics[18]	X	X			www.rii.com
Sigma-Genosys[1,2,8,11,12,13,18]	X				www.genosys.com
Super Array Inc.[1,2,18]	X				www.superarray.com
Takara[1,2,4, 8,17,18]		X			www.takara.co.jp/english/bio_e

Notes:

[1]Human, [2]Mouse, [3]Rat, [4]*Arabidopsis*, [5]*Drosophila*, [6]*Saccharomyces cerevisiae*, [7]HIV, [8]*Escherichia coli*, [9]*Candida albicans*, [10]*Staphylococcus aureus*, [11]*Bacillus subtilis*, [12]*Helicobacter pylori*, [13]*Campylobacter jejuni*, [14]*Streptomyces lividans*, [15]*Streptococcus pneumoniae*, [16]*Neisseria meningitidis*, [17]Cyanobacteria, [18]Custom.

synthesized DNA mircoarrays, and bring this activity into individual research laboratories.

CombiMatrix (Snoqualmie, WA) and Nanogen (San Diego, CA) are developing electrical addressing systems for the manufacture of DNA arrays on semiconductor chips. The CombiMatrix method involves attaching each addressable site on the chip to an electrical conduit (electrode) applied over a layer of porous material. Each DNA probe is synthesized one base at a time by flooding the porous layer with a nucleoside and activating each electrode where a new base is to be added. Once activated, the electrode causes an electrochemical reaction to occur which produces chemicals that react with the existing nucleotides, or chains of DNA, at that site for bonding to the probe site or to the next nucleotide base. At present, CombiMatrix has produced DNA arrays with 100 μm features that possess 1024 test sites within less than a square centimeter. Researchers at CombiMatrix believe that by using a standard 0.25-μm semiconductor fabrication process, they can produce a biological array processor with over 1 000 000 sites per square centimeter.

Nanogen uses a similar process to attach pre-synthesized oligonucleotides to electronically addressable sites on a semiconductor chip. To date, Nanogen has only produced a 99 probe array suitable for forensic and diagnostic purposes; however, Nanogen's researchers anticipate electronic arrays with thousands of addresses for genomics applications.

Pre-synthesized DNA arrays

The method of attaching pre-synthesized DNA probes (usually 100–5000 bases long) to a solid surface such as glass (or nylon filter) supports was conceived 25 years ago by Ed Southern and more recently popularized by the Patrick O. Brown laboratory at Stanford University. While the early manufacturing methods for miniaturized DNA arrays using *in situ* probe synthesis required sophisticated and expensive robotic equipment, the glass slide DNA array manufacturing methods of Brown made DNA arrays affordable for academic research laboratories. As early as 1996 the Brown laboratory published step-by-step plans for the construction of a robotic DNA arrayer on the internet. Since that time, many commercial DNA arrayers have become available. Besides the commercially produced Affymetric GeneChips™, these Brown-type glass slide DNA arrays are currently the most popular format for gene expression profiling experiments.

The Brown method for printing glass slide DNA arrays involves the robotic spotting of small volumes (in the nanoliter to picoliter range) of a DNA probe sample onto a $25 \times 76 \times 1$ mm glass slide surface previously coated with poly-lysine or poly-amine for electrostatic adsorption of the DNA probes onto the slide. Depending upon the pin type and the exact printing technology employed, 200 to 10000 spots ranging in size from 500 to 75 μm can be spotted in a 1-cm^2 area. Many public and private research institutions in the USA and abroad have developed core facilities for the in-house manufacture of custom glass slide DNA arrays. Detailed discussions of the instrumentation and methods for printing glass slide DNA arrays can be found in a book edited by Mark Schena titled *Microarray Biochip Technology* [3].

Several additional methods for attaching pre-synthesized DNA probes to solid surfaces for the manufacture of DNA arrays have been developed. For example, the electronic methods of CombiMatrix and Nanogen and the piezoelectric methods of Agilent Technologies, described earlier, can also be used to attach pre-synthesized DNA probes to glass slides. In addition to these methods, Interactiva (Ulm, Germany) has developed a method for the attachment of DNA probes to a 0.1 μm layer of 24-karat gold overlaid with a hydrophobic teflon covering containing an array of 50 μm wells. A self-assembling monolayer of thio alkanes coupled to streptavidin is adsorbed to the gold surfaces in each well. This streptavidin surface serves as an anchor for the deposition of biotin-labeled DNA probes.

Corning (Corning, NY) has developed a novel manufacturing process that enables the high speed, high capacity production of arrays. Drawing upon their experience in glass and plastics manufacturing, process engineering, and surface technologies, they have developed a method to simultaneously array thousands of spots onto specially formulated glass slides in a massively parallel and reproducible manner and at a high density. This process is based on Corning's ability to construct a honeycomb-like glass structure with 1024 individual cells. Each cell is loaded with a unique DNA probe and the contents of the honeycomb structure are simultaneously printed onto the surface of a glass slide. Since more than a thousand probes are deposited onto each slide at the same time this process is much faster than current methods, and it allows the simultaneous printing of thousands of slides per manufacturing run. Unfortunately, Corning abandoned this array manufacturing process early in 2002.

Filter-based DNA arrays

Although solid support matrices such as glass offer many advantages for high-throughput processing, nylon filter-based arrays continue to be a popular format. The primary reasons for this appear to be that gene expression profiling with nylon filters is based on standard southern blotting protocols familiar to molecular biologists, and because equipment to perform filter hybridizations with ^{33}P-labeled cDNA targets and for data acquisition, such as a phosphorimager, are available at most research institutions.

Although, primarily because of health risks, non-radioactive labeling is preferred in most settings, radioactive labeling of the targets for hybridization to nylon filter arrays offers the advantage of greater sensitivity compared to fluorescently labeled targets, and intensity measurements linear over a four to five log range are achieved with radio-labeled targets; whereas, linear ranges of only three logs are typically observed with fluorescently labeled targets. Additional advantages of nylon filter arrays are that many types are available on the commercial market (Table 2.1), and that there is a cost advantage inherent in the fact that nylon filters can be stripped and reused several times without significant deterioration [4].

Because of the porous nature of nylon filters, they are not amenable to miniaturization. However, they are suited for gene expression profiling studies in organisms with small genome sizes such as bacteria, or for the production of custom arrays containing DNA probes for a functional subset of an organism's genes. For example, Sigma-Genosys (Woodland, TX) produces a nylon filter DNA array for gene expression profiling in *E. coli*. This filter array, which measures 11×21 cm, contains over 18000 addresses spotted, in duplicate, with full-length PCR products of each of the 4290 *E. coli* open reading frames (ORFs). In addition to *E.coli* arrays, Sigma-Genosys also produces high-quality nylon filter arrays containing duplicate full-length PCR products of all of the ORFs of the bacteria *Bacillus subtilis* and *Helicobacter pylori* and additional genomic arrays of other bacteria are being developed. They and others also provide a nylon filter array containing human cytokine genes and an array containing human apoptosis genes.

Several other companies also provide nylon filter arrays similarly spotted with DNA probe sets of related human genes (Table 2.1). For example Clonetech (Palo Alto, CA) produces nearly 30 nylon filter DNA arrays containing from 200 to 1200 DNA probes for functionally related sets of human, mouse, or rat genes. Research Genetics (Minneapolis, MN) sells filters with probes for functionally related human, mouse, and rat genes as well as the complete genome of *Saccharomyces cerevisiae* (6144 probes) contained on two nylon membrane filters. New products from these and other commercial suppliers are appearing at a rapid rate.

Non-conventional gene expression profiling technologies

Lynx (Hayward, CA) has developed yet another strategy for identifying and isolating differentially expressed genes between two cell types. Their method employs a "fluid array." This method involves hybridizing two probes prepared separately, one from each of the samples to be compared, with a population of micro-beads, each carrying many copies of a single DNA fragment or gene derived from either of the samples. Because each probe is labeled with a different fluorescent marker, genes or fragments that are under- or overrepresented in either sample are readily separated in a fluorescence activated cell sorter. Genes or fragments of interest can be recovered and cloned for further study.

The Lynx approach has been extended by an intriguing bead-based fiber optic DNA array format suggested by David R. Walt of Tufts University. Since this format is a dramatic departure from other array platforms it warrants some comment. First, an array of thousands (typically 5000–50000) of individual optical fibers each 3 to 7 μm in diameter are fused into a single bundle. Next, an agent such as hydrofluoric acid is used to etch 2 μm wells (the size of microspheres) into the ends of each fiber, and microspheres each carrying a different DNA probe are inserted into the wells. The microspheres can contain oligonucleotide probes built up base by base using conventional phosphoramidite chemistry, or preformed oligonucleotides can be added directly to surface-activated microspheres. Fluorescent signals generated by probe hybridization to target DNA molecules on the microspheres are transmitted through the fiber to a photo-detection system.

Since the resulting array is randomly distributed, any probe sequence can be positioned in any given well. Therefore, a strategy must be available for registering each array. Walt suggests several methods to accomplish this task. For example, each different type of microsphere can be tagged with a unique combination of fluorescent dyes either before or after probe attachment. This "optical bar code" is a combination of fluorescent dyes with different excitation and emission wavelengths and intensities that allow each bead to be independently identified. These optically bar-coded arrays can be decoded with conventional signal processing software by collecting a series of fluorescence signals at different excitation and emission wavelengths and analyzing the relative intensities emitted from each bead.

Because of the necessity to insure that each array contain at least one bead for each probe, replicates of each bead must be present in each array. However, this redundancy provides two advantages. First, a voting scheme in which replicates must agree can be used to evaluate false positives and false negatives. Second, redundancy enhances sensitivity. Since signal-to-noise ratios scale as the square root of the number of identical sensing elements, sensitivity can be enhanced by looking at all the identical probe microspheres in the array. A disadvantage for the creation of fiber optic arrays is that each bead-based oligonucleotide probe must be synthesized individually as opposed to the combinatorial synthesis approaches available with light-directed or ink-jet techniques. On the other hand, a single synthesis can produce enough beads for thousands of arrays.

An advantage of the fiber optic array is its sensitivity. Walt argues that the small size of individual beads ensures a high local concentration when only a few DNA target copies are bound. For example, a 3 μm bead occupies a volume of only a few femtoliters; therefore, when only 1000 labeled target molecules hybridize to the probe microsphere, a local target concentration near 1 mmol is achieved – a relatively easy concentration for fluorescent dye to detect. Walt points out that this local concentration of fluorescent probe combined with the signal-to-noise enhancements from redundant array elements will enable detection limits of femtomolar concentrations and absolute detection limits of zeptomoles (10^{-21} moles) of DNA.

A limitation to this technology has been the fact that current fluorophores fade quickly, have a limited color range, and their overlapping absorption spectra make their combined use impractical for high-level multiplexing applications. However, researchers from the University of Indiana have recently described a sophisticated optical labeling system that relies on the unique properties of luminescent quantum dots (QDs) – light-emitting nanocrystals. Like conventional fluorophores, QDs absorb light and emit it at a different wavelength. However, QDs are brighter than conventional fluorophores, and they are resistant to photobleaching. QDs of different sizes emit different colors that can be excited by the light of a single wavelength, and unlike ordinary fluorophores, the emission spectra of individual QDs are very narrow. Thus, QDs can be embedded in polymer beads at precisely controlled ratios to produce a rainbow of colors. It appears that QD nanocrystals embedded in microspheres might be ideal for the bar-coding applications envisioned by Walt and others.

REFERENCES

1. Fodor, S. P., Rava, R. P., Huang, X. C., Pease, A. C., Holmes, C. P., and Adams, C. L. Multiplexed biochemical assays with biological chips. 1993. *Nature* 364:555–556.
2. Singh-Gasson, S., Green, R. D., Wue, Y., Nelson, C., Blattner, F., Sussman, M. R. and Cerrine, F. Maskless fabrication of light-directed oligonucleotide microarrays using a digital micromirror array. 1999. *Nature Biotechnology* 17:974–978.
3. Schena, M. (ect.) *Microarray Biochip Technology*. 2000. Eaton Publishing Co., Natick, MA.
4. Arfin, S. M., Long, A. D., Ito, E., Riehle, M. M., Paegle, E. S., and Hatfield, G. W. Global gene expression profiling in *Escherichia coli* K12: the effects of integration host factor. 2000. *Journal of Biological Chemistry* 275:29672–29684.

3

DNA array readout methods

Once a DNA array experiment has been designed and executed the data must be extracted and analyzed. That is, the signal from each address on the array must be measured and some method for determining and subtracting the background signal must be employed. However, because there are many different DNA array formats and platforms, and because hybridization signals can be generated with fluorescent- or radioactive-labeled targets, no single DNA array readout device is suitable for all purposes. Furthermore, many instruments with different advantages and disadvantages for different types of array formats are available. Therefore, since accurate data acquisition is a critical step of any array experiment, careful attention must be paid to the selection of data acquisition equipment.

Reading data from a fluorescent signal

All arrays that emit a fluorescent signal must be read with an instrument that provides a fluorescence excitation energy source and an efficient detector for the light emitted from the fluorophore incorporated into the target. Currently, the fluorophores most commonly used for incorporation into cDNA targets are Cy3 and Cy5. These are cyanine dyes commercially available as dUTP or dCTP conjugates. Cy3 is an orange dye with a light absorption maximum at 550 nm and an emission maximum at 581 nm. Cy5 is a far-red dye with a light absorption maximum at 649 nm and an emission maximum at 670 nm. All commercial scanners for reading fluorescent array signals use lasers and filters that provide an excitation light source at a single wavelength near the absorption maxima of these and other commonly used dyes. Lasers eliminate pollution of fluorescence emissions at nearby wavelengths that would occur with a light source of multiple wavelengths. Defined laser emission wavelengths and filters also permit the use and efficient detection of fluorescent emissions from two or more fluorophores in a single experiment. This permits the common practice of combining cDNA targets prepared from a reference and an experimental condition, one containing Cy3-labeled and the other Cy5-labeled nucleotides, for analysis on a single array. Of course, instruments for this application must be equipped with multiple laser sources, one for each fluorophore used.

In addition to providing an appropriate light source for fluorophore excitation, attention must be paid to the efficiency and accuracy of fluorescence emission measurements. Efficiency is an issue because the amount of light emitted from a fluorophore is generally orders of magnitudes (as much as 1 000 000-fold) weaker than the intensity of the light required to excite the fluorophore. This problem is further exacerbated by the fact that fluorophores emit light in all directions, not just toward the light detector system. Thus, some optical method to collect as much of the emitted light and eliminate as much of the excitation light as possible is required. This is accomplished with an objective lens that captures the emitted light and focuses it toward the detector. However, it is not possible to collect this light simultaneously from all directions. The best that can be accomplished is to collect that light that is emitted from the hemisphere of the fluorescence directed toward the detector. The efficiency of an objective lens in accomplishing this task is expressed as its numerical aperture. If it were possible to collect all of the light in a hemisphere, the objective would have a numerical aperture of 1.0. In reality, objective lenses in array readers have numerical apertures ranging from 0.5 to 0.9. Obviously, the numerical aperture of the objective lens directly affects an instrument's sensitivity. Therefore, since instrument manufacturers often do not list this information in their product description, it is something that a purchaser should ask for.

All organic molecules fluoresce. Any contamination on an array surface (commonly glass slides), including the coated surface of the slide itself, will produce background fluorescence that can seriously compromise the signal-to-noise ratio. The answer to this problem has been addressed by limiting the focus of the laser to a small field of view in three-dimensional space centered in the DNA sample on the array. This is accomplished by a confocal method comprising two lenses in series producing two reciprocal focal points, one focused on the sample and the other focused at a pinhole for transmission of the light signal to the detector. The beauty of this system is that the pinhole permits only the light from a very narrow depth of focus of the objective lens to be passed through to the detector (Figure 3.1). This allows very fine discrimination of noise and signal.

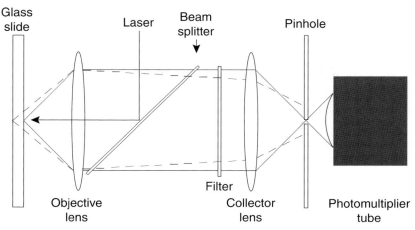

Figure 3.1. The principle of confocal scanning laser microscopy. Confocal systems image only a small point (pixel) in three-dimensional space. This is achieved by employing a laser as an illumination source and a small aperture in front of the detector, which usually is a photomultiplier tube. A schematic of a simplified confocal optical path is shown above. The laser beam enters from above and is reflected by the beam splitter (a dichroic or multichroic filter that reflects light of short wavelengths and transmits light of longer wavelengths). The laser beam is focused to a spot on the surface of the glass slide by an objective lens and excites fluorophores in the focal plane of the objective. The fluorescent emissions (and the reflected laser beam rays) are collected by the objective and collimated into parallel rays (solid lines). Most of the reflected laser rays (>95%) are reflected back toward the laser source by the beam splitter. The remaining laser rays are excluded by a downstream filter specific for the emission wavelength of the fluorophore. The parallel rays of the fluorescent light received from the focal plane of the objective lens are focused on a pinhole in a detector plate and passed through to the photomultiplier tube. The ray paths indicated with dashed lines shows how light from out-of-focus objects such as reflections and emissions from the second surface of the glass slide are eliminated by the design of the optical system. This out-of-focus fluorescent light takes a different path through the objective lens, the beam splitter, and the emission filter. As a result, it is not focused into the detector pinhole and, therefore only a small portion of this polluted light is passed on to the photomultiplier tube. Since confocal microscopy images only a small point in three-dimensional space proportional to the aperture of the pinhole in the detector plate, complete images must be obtained by scanning and digital reconstruction.

The depth of focus for most commercial laser-based array readers is 3 μm, less than half the thickness of the DNA sample on the slide surface. However, since the diameters of the spots printed on glass slide DNA arrays range from 25 to 100 μm, this means that at this 3 μm (1 pixel) resolution slides must be scanned to collect all of the data even from a single spot. At this resolution, even the highest density arrays can be accurately scanned. Scanning at this resolution, however, requires several minutes. For this reason, manufacturers of confocal laser scanners provide the option to scan at lower resolutions ranging down to 50 μm. This is useful for rapid pre-scanning of arrays for sensitivity and scanning alignment adjustments.

There are three ways to scan an array – either move the array or move the laser, or both. In general systems that move the array under a fixed laser source are preferred because of their increased light-gathering efficiency.

Another consideration for a confocal laser scanner is the detector system and its associated software for data aquisition. Most commercially available instruments use photomultiplier tubes with amplifiers that allow for high voltage adjustments that can provide up to a 10000-fold sensitivity range, sufficient for accommodating the signal variability and intensity ranges of most array experiments. In addition, the power of the laser can be adjusted. The greater the intensity of the laser beam the more fluorescence; however, this is partially offset by the short time during scanning that each pixel is illuminated. The optimal strength of laser intensity is further dictated by the fluorophore since excessive excitation light can damage the fluorophore and decease the signal-to-noise ratio (photo bleaching). Thus, efforts are under way to develop new fluorophores that can accept more excitation energy, and produce more fluorescence.

Some commercially available scanners are equipped with a charge-coupled device (CCD) to detect photons rather than a photomultiplier tube to create an array image. The major drawbacks of CCD systems are the time required for image acquisition and the limited resolution of the image obtained. CCD systems use white light to excite the fluorophores on the array. Since this results in the release of many fewer photons than a focused high-intensity laser source, more time is required to generate an image of acceptable quality. This extended integration time results in higher backgrounds due the generation of dark-current noise. Also, even at a 10 μm resolution the largest CCD chips (approx. 1600×1200 pixels) available at an acceptable price are too small to acquire a single image from a 25×76 mm glass slide array. Thus, laser scanners equipped with photomultiplier tubes are the instruments of choice for most researchers.

All scanners on the market today are controlled through a computer interface to set instrument parameters and retrieve array data. In addition, each manufacturer includes software bundles to filter and process these data, often in proprietary ways that may not fit the experimental designs of every researcher. Thus, given the myriad of ways that individual researchers might wish to process and analyze array data and the rapid advances occurring in this area, it is important to ascertain that the instrument's data aquisition software fits the user's needs and allows retrieval of primary unprocessed data.

At present, there are several major suppliers of scanners for DNA arrays. The essential features of the instruments offered by these companies are described in Table 3.1.

Reading data from a radioactive signal

Historically, nucleic acids have been labeled with beta-emitting radioisotopes such as ^{32}P or ^{33}P incorporated into the α or γ phosphate position of nucleotide triphosphate molecules and visualized by exposure to X-ray film. While this is an efficient and proven labeling and imaging method routinely used in molecular biology laboratories, it is not suitable for small format, high-density DNA array platforms with DNA probes printed at 100 to 300 μm intervals on glass slides or 20 to 24 μm intervals on Affymetrix GeneChips™. This is because the omni-directional radioactive emissions from targets hybridized to one spot on the array can contaminate the signals obtained from neighboring probe sites. Furthermore, imaging and quantitation of arrays with X-ray film is excluded because the grain density is too large and the linear range of the film (<100-fold) is far less than the linear range of signal intensities (>10000-fold) obtained from array experiments. However, larger format arrays are feasible, and the signals from these arrays can be accurately measured by commercially available phosphorimaging instruments with a linear range over 100000-fold.

Figure 3.2 shows a phosphorimage of a large-format Sigma-Genosys Panorama™ *E. coli* Gene Array printed on a 11×21 cm nylon membrane. This array contains over 18000 addresses. The size of each probe spot is 1 mm and they are centered at 1 mm intervals. Quantitative data can be extracted from DNA arrays of this type with any of a number of phosphorimaging instruments with a scanning resolution of at least 50 μm.

Phosphorimaging is based on a phenomenon known as photo-stimulated luminescence (PSL), a still poorly understood phenomenon said to have been discovered by Henri Becquerel in the mid eighteen hundreds. To understand the basic principle of PSL it is necessary to know that fluorescence differs from phosphoresence in that excited electrons of fluorescent atoms immediately return to their base state and emit fluorescent light, whereas excited electrons of phosphorescent atoms only slowly return to their base state and continue to emit light for a while after the stimulation stops. The PSL phenomenon is based on a substance that

Table 3.1. *Commercial sources for confocal laser scanners*

Company	Web address	Model	Number of Lasers	Scan area	Sensitivity
Affymetrix	www.affy metrix.com	428 Array Scanner	1	22×75 mm	<1 Cy3 molecule/mm at greater than 3.5 S/N
Alpha Innotech Corp.	www.alpha innotech.com	Alpha Array 7000	White lights	16×22 mm	0.06 fluor/ μm²
Amersham/ Pharmacia Biotech[a]	www.mdyn.com	Gene Pix 4000A	2	22×73 mm	0.1 molecule fluor/μm²
Applied Precision	www.applied precision.com	array WoRx	White light	22×60 mm	0.1 molecule FAM/μm²
GeneFocus	www.genefocus.com	ORF DNA scope	1	1×1 mm to 22×22 mm	0.1 molecule fluor/μm²
Genomic Solutions	www.genomic solutions.com	Gene TAC LS IV	4		0.1 molecule fluor/μm²
Nanogen	www.nanogen.com	Nano chip Reader	2	2×2 mm	
Packard	www.packard inst.com	Scan Array 4000	2	22×73 mm	<0.1 molecule fluor/μm²
Packard	www.packard inst.com	Scan Array 4000XL	3	22×73 mm	<0.1 molecule fluor/μm²
Packard	www.packard inst.com	Scan Array 5000	4	22×73 mm	<0.1 molecule fluor/μm²
Packard	www.packard inst.com	Scan Array 5000XL	5	22×73 mm	<0.1 molecule fluor/μm²
Virtek	www.virtek.ca	Virtek Chip Reader	2	22×65 mm	0.1 molecule fluor/μm²
Vysis	www.vysis.com	Geno Sensor Reader	White light	<0.5 cm²	0.1 molecule fluor/μm²

Notes:
[a] Not confocal.
[b] PMT, photomultiplier tube; CCD, charge-coupled device

Pixel resolution	Filters	Mechanism of scanning	Scan speed	Supported dyes	Source type
3, 6, 9, 12, 24 μm	(Up to 6) Std. 551, 570, 665 nm	PMT	4 min	Fluorescein and phyco-erythrin	Laser argon-ion 488 nm
3–30 μm	Various	CCD	15 s	Various	Nova Ray Light Mgt. System
5–100 μm	None	Dual PMT	<5 min	Cy dyes	Laser diodes 532, 635 nm
5–50 μm	330–780 nm (4 filters)	CCD	4 min	Cy3 and Cy5	Halide 380–800 nm
16 bits/ pixel	Various	PMT	30 s to 15 min	Cy3 and Cy5 Others optional	Laser
1–100μm	Various	PMT	2 min	Various fluorescent dyes	Laser
	532 nm 635 nm	PMT		Cy3 and Cy5	Lasers 532, 635 nm
5, 10, 20, 30, 50 μm	488–633 nm (6 filters)	PMT	<5 min	Various dyes	Confocal lasers
5, 10, 20, 30, 50 μm	543, 594, 633 nm (7 filters)	PMT	<5 min	Various dyes	Confocal lasers
5, 10, 20, 30, 50 microns	500–700 nm (10 filters)	PMT	<5 min	Various dyes	Confocal lasers
5, 10, 20, 30, 50 μm	488, 513, 543, 594, 633 nm (11 filters)	PMT	<5 min	Various dyes	Confocal lasers
10 μm 20 μm	532, 635 nm	PMT	3 min	Cy3 and Cy5	Laser diodes
	(3 filters)	CCD	1 min	Various dyes	Xenon

1 2 3 4 5 6 7 8 9 10 11 12 13 14 15 16 17 18 19 20 21 22 23 24

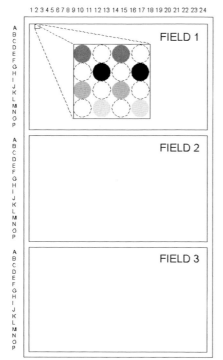
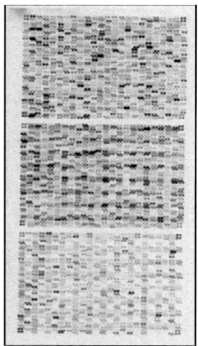

Figure 3.2. Sigma-Genosys Panorama™ *E.coli* Gene Array. Each *E. coli* array contains three fields divided into 24×16 (384) primary grids (1–24×A-P). Each primary grid contains 16 secondary grids separated by 1 mm. Four different full-length ORF probes are spotted in duplicate in every other secondary grid as shown in the blowup of Field 1, A2. The remaining eight secondary grids of each primary grid are blank. In the third field, some primary grids do not contain any ORFs.

differs from phosphorescence in that the primary stimulation results in the trapping of an excited electron in a higher energy state. This electron is not released to return to its base state to release fluorescent light until it is excited by a second stimulation of light, usually from a laser, having a longer wavelength than the primary stimulation source, for example radiation.

A phosphorimager screen uses BaFBr:Eu^{+2} crystals to store radiation information and releases it at a later time. Phosphorimaging screens contain a layer of these crystals (grain size approx. 5 μm) contained between a support and a transparent protective layer. When a europium electron is excited by a primary energy source (radiation) it is trapped in the crystal lattice structure producing a stable BaFBr:Eu^{+3} crystal with a

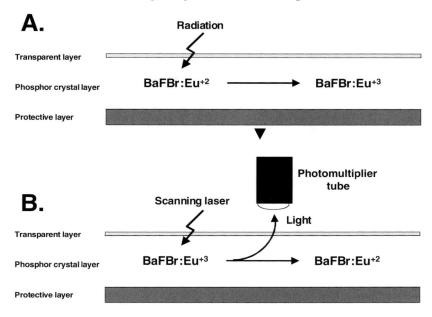

Figure 3.3. The principle of photo-stimulated luminescence.

visible light color center. This structure is stable until the crystal is excited a second time with laser light absorbed by the color center. This releases the trapped electron that emits fluorescent light captured by a photomultiplier tube during a laser scan of the screen. Since all of the crystals are not returned to the $BaFBr:Eu^{+2}$ state during the scan, the screen is "erased" for subsequent use by flashing with visible light. The basic principle of PLS is schematically illustrated in Figure 3.3.

Thus, when a DNA array hybridized with radioactively labeled targets is placed on the phosphorimaging screen in the cassette, trapped electrons are stored in the $BaFBr:Eu^{+3}$ crystals at a rate proportional to the intensity of the radioactive signal. After an appropriate exposure time (often 12–48 hours) the DNA array is removed and the phosphorimaging screen is scanned with a laser and a photo-optical detection system containing a photomultiplier tube that measures and records the emitted light. This information is digitized and reconstructed into an image of the type shown in Figure 3.2.

Since the scanning surfaces accommodated by phosphorimagers are much larger than those of glass slide arrays scanned with confocal laser scanners, time constraints dictate that they cannot be scanned at the same resolution. Therefore, to accommodate this time constraint, the highest scanning resolution for phosphorimaging instruments is 20 μm; however, as scanning mechanics improve, scanning resolutions approaching the 5 μm size of the phosphor crystal are anticipated.

Some newer storage phosphorimaging instruments come equipped with multiple scanning laser sources and filters for the excitation of many commonly used fluorophores. For example, fluorescent dye combinations such as Cy3 and Cy5 can be imaged with these instruments. Nevertheless, while these detection methods are suitable for many applications, they do not offer the signal-to-noise and scanning resolution advantages necessary for high-density DNA arrays that are provided by confocal laser scanning instruments. The features of several currently available phosphorimagers are described in Table 3.2.

Table 3.2. *Commercial sources for phosphorimagers*

Name	Web address	Model	Resolution	Scan area	Detects	Linear dynamic range
BioRad	www.bio-rad.com	*Molecular Imager FX*	50,100, 200, 800 μm	35×43 cm	Radioisotopes Fluoresence Chemiluminesce	$1:10^5$
Molecular Dynamics	www.mdyn.com	*Phosphor imager SI*	50, 100, 200 μm	35×43 cm	Radioisotopes	$1:10^4$
		Storm	50,100, 200 μm	35×43 cm	Radioisotopes Fluoresence Chemiluminesce	$1:10^5$
		Typhoon	50,100, 200 μm	35×43 cm	Radioisotopes Fluoresence Chemiluminescence	$1:10^5$
Fuji	www.fujimed.com	*BAS5000*	25 and 50 μm	20×25 cm	Radioisotopes	$1:10^4–10^5$
Packard	www.packardinst.com	*Cyclone*	50 μm	35×43 cm	Radioisotopes	$1:10^5$

4

Gene expression profiling experiments: Problems, pitfalls, and solutions

In previous chapters, we have discussed the formats, methods of manufacture, and data aquisition instruments required for gene expression profiling with DNA arrays. In this chapter, we consider issues such as: heterogeneities encountered among experimental samples, isolation procedures for non-polyadenylated and polyadenylated RNA from bacteria and higher organisms; advantages and disadvantages of different target preparation methods; and general problems encountered during the execution of such experiments. Throughout this chapter we focus on ways to minimize experimental errors. In particular, we point out the pitfalls of current methods for the preparation of targets from polyadenylated RNA and discuss alternative methods of target synthesis from total RNA preparations from eukaryotic cells based on methods developed for the synthesis of bacterial targets. Regardless of the fact that we model many of our discussions around bacterial systems, it should be emphasized that the lessons of this chapter are applicable to gene expression profiling experiments in all organisms.

Primary sources of experimental and biological variation

Differences among samples

It is, of course, desirable to minimize extraneous biological and experimental variables as much as possible when analyzing gene expression profiles obtained under two defined experimental conditions such as a temporal or treatment gradient, or between two different cell sample types or genotypes [1, 2, 3, 4]. In Chapter 7, we compare the gene expression profiles between two genotypes, lrp^+ and a lrp^- strains of *E. coli*. To minimize differences between the genotypes of these cells, care was taken to insure that both strains were isogenic; that is, that they contained identical genetic back-grounds except for the single structural gene for the Lrp protein. This was accomplished by using a single wild-type *E. coli* K12 strain for the con-struction of the two isogenic strains. First, the natural promoter-regulatory region of the *lacZYA* operon in the chromosome was replaced with the pro-moter of the *ilvGMEDA* operon (known to be regulated by Lrp). This pro-duced a strain (IH-G2490; *ilv*P$_G$::*lacZYA, lrp$^+$*) in which the known effects of Lrp on *ilvGMEDA* operon expression could be easily monitored by enzymatic assay of the gene product of the *lacZ* gene, β-galactosidase. Next, the Lrp structural gene was deleted from this strain to produce the otherwise isogenic strain (IH-G2491; *ilv*P$_G$::*lacZYA, lrp$^-$*) [5].

The ability to control this source of biological variation in a model organism such as *E. coli* with an easily manipulated genetic system is an obvious advantage for gene expression profiling experiments. However, most systems are not as easily controlled. For example, human samples obtained from biopsy materials will not only differ in genotype but also in cell types. Nevertheless, the experimenter should strive to reduce this source of biological variability as much as possible. For example, laser-capture techniques for the isolation of single cells from animal and human tissues for isolation and amplification of RNA samples that address this problem are being developed [6].

An additional source of biological variation in experiments comparing the gene profiles of two cell types comes from the conditions under which the cells are cultured. In this regard we have recommended that standard cell-specific media should be adopted for the growth of cells queried by DNA array experiments [2]. While this is not possible in every case, many experimental conditions such as the comparison of two different genotypes of the same cell line can be standardized. The adoption of such medium standards would greatly reduce experimental variations and facilitate the cross-comparison of experimental data obtained from different experi-ments and/or different experimenters. For *E. coli*, Neidhardt *et al.* [7] have performed extensive studies concerning the conditions required for main-taining cells in a steady state of balanced growth. Their studies have defined a synthetic glucose-minimal salts medium (glucose-minimal MOPS) that minimizes medium changes during the logarithmic growth phase. The experiments described in Chapter 7 were performed with cells grown in this medium. Similar studies have described defined media for the growth of many eukaryotic cell lines that should be agreed upon by researchers and used when experimental conditions allow.

RNA isolation procedures

Another large, in our experience the largest, source of error in data from DNA array experiments comes from the biological variations in individual mRNA expression levels in different cell populations, even when care is taken to culture cells under "identical" conditions [2]. This problem is exacerbated if extreme care in the treatment and handling of the RNA is not taken during the extraction of the RNA from the cell and its subsequent processing. For example, it is often reported that the cells to be analyzed are harvested by centrifugation and frozen for RNA extraction at a later time. It is important to consider the effects of these experimental manipulations on gene expression and mRNA stability. If the cells encounter a temperature shift during the centrifugation step, even for a short time, this could cause a change in the gene expression profiles due to the consequences of temperature stress. If the cells are centrifuged in a buffer with even small differences in osmolarity from the growth medium, this could cause a change in the gene expression profiles due to the consequences of osmotic stress. Also, removal of essential nutrients during the centrifugation period could cause significant metabolic perturbations that would result in changes in gene expression profiles. Each of these and other experimentally caused gene expression changes will confound the interpretation of the experiment.

These are not easy variables to control. Therefore, the best strategy is to harvest the RNA as quickly as possible under conditions that "freeze" it at the same levels that it occurs in the cell population at the time of sampling. For example, in the experiments described in Chapter 7, total RNA was isolated from cells during balanced growth in a defined glucose-minimal MOPS medium during the mid-logarithmic phase of growth at a cell density of $OD_{600} = 0.6$. Avoidance of temperature stress was attempted by using a pipette equilibrated to the same temperature as the cell growth medium to rapidly transfer cells from the growth flask into a boiling solution of a cell lysis buffer containing an ionic detergent, sodium docecyl sulfate (SDS), described in Appendix A. This procedure instantly disrupts the cells and inhibits any changes in mRNA levels due to degradation by endogenous RNase activities. Additionally, all solutions and glassware used in this and subsequent steps are treated with an RNase inhibitor, diethylpyrocarbonate (DEPC). Rapid RNA isolation methods designed to minimize changes in gene expression profiles, and inhibit RNase activities, during RNA isolation from eukaryotic cells and tissues have also been described. For example, Ambion Inc. (Austin, TX) produces a product called RNA *later*™. This is a storage reagent that inactivates RNase activities and stabilizes cellular RNA in intact unfrozen cells and tissue samples. Tissue pieces or cells can be submerged in RNA *later*™ for storage without compromising the quality or quantity of RNA obtained after subsequent RNA isolation.

Even employing these experimental conditions to "freeze" the extracted RNAs at the levels present in the growing cells, comparisons of DNA array data of experimental replicates obtained with RNA preparations from the same cells, growing under the same conditions, reveal a significant degree of variability in gene expression levels. To reduce this variation even further, RNA preparations from three separate extractions from multiple cultures of cells growing under the same experimental conditions can be purified, labeled, and pooled prior to hybridization to an array. This practice tends to average out biological variations in gene expression levels due to, for instance, subtle differences in growth conditions and labeling efficiencies, and significantly reduces the false positive levels observed in gene expression profiling experiments (see Chapter 7).

Special considerations for gene expression profiling in bacteria

The first step in any gene expression profiling experiment, whether it is with bacteria or higher eukaryotic organisms, is to isolate mRNA for the production of labeled targets to hybridize to DNA arrays. With eukaryotic organisms that contain 3′ polyadenylated mRNA, it is simply purified away from bulk cellular RNA with oligo(dT) chromatography columns. However, unlike mRNA from eukaryotic cells, the vast majority of mRNA in bacteria is not polyadenylated. Thus, it was necessary to develop some other method to isolate bacterial mRNA. Otherwise, since mRNA comprises less than 10 percent of the total RNA in a bacterial cell, and indeed the cells of all organisms, it was feared that targets produced from total RNA preparations would produce unacceptable backgrounds because those targets derived from ribosomal RNA and other non-mRNA species would cross-hybridize to array probes.

An early solution proposed to obtain labeled targets specifically derived from mRNA in total RNA preparations from bacteria was to synthetically prepare a collection of 3′-oligonucleotide primers specific for the 3′ ends of each bacterial open reading frame (ORF). These primers then could be used for DNA polymerase-mediated primer-directed synthesis of radioactive- or fluorescent-labeled cDNA targets for hybridization to DNA arrays containing ORF probes in much the same way that oligo(dT) primers are used for target synthesis with poly(A) mRNA from eukaryotes. In fact, the first commercially available *E. coli* DNA arrays produced by Sigma-Genosys were offered with such a set of oligonucleotide primers. However, while several gene expression profiling experiments using these message-specific oligonucleotide primers appeared in the literature, it was soon realized that they did not produce acceptable results.

A comparison of the use of targets prepared with message-specific primers and targets prepared with random hexamer primers from total RNA preparation for use with pre-synthesized DNA arrays

The alternative to the use of 3′ message-specific primers for the synthesis of bacterial targets was to prepare targets from total RNA preparations with random primers. Early in 2000 we performed a systematic set of experiments to compare these alternatives [2]. For these experiments, we used Sigma-Genosys nylon filter DNA arrays spotted in duplicate with full-length PCR products of each of the 4290 *E. coli* ORFs (Figure 3.2). We performed each experiment in duplicate and replicated each duplicate experiment four times. This experimental design, which results in four measurements of each target in each of four independent experiments, is diagrammed in Figure 4.1.

In one case, we carried out this series of four independent experiments with ³³P-labeled cDNA targets prepared by primer extension of total RNA preparations directed by random hexamer oligonucleotides. In the other case, we performed the same series of experiments with ³³P-labelled cDNA targets prepared by primer extension of total RNA preparations directed by a complete set 3′ message-specific primers for each *E. coli* ORF. We found that at the high stringency hybridization conditions that can be used with arrays containing full-length ORF probes low backgrounds are observed (Figure 3.2), and that cDNA targets prepared with the random hexamer oligonucleotides detected an average expression of 2592 genes with at least two out of four, non-zero, background-subtracted measurements for both the control and experimental samples. However, filter hybridization with the cDNA targets prepared with the message-specific primers detected an average expression of only 1760 genes with at least two out of four non-zero, background-subtracted measurements for both samples. Thus, we detected one-third more genes with the random-hexamer-derived targets than with the message-specific-derived targets.[1] Equally disturbing, we observed that while some genes of a given

---

[1]   Under the stringent hybridization conditions employed in this experiment (Appendix A), no hybridization of ³³P-labeled targets prepared with random hexamers from purified ribosomal RNA was detected [2].

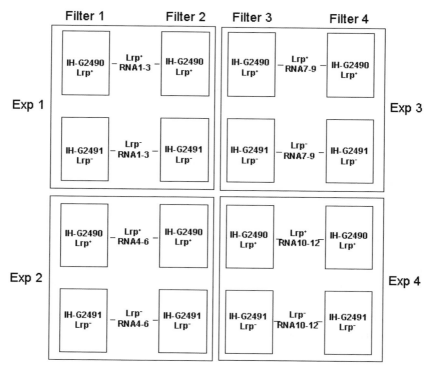

Figure 4.1. Experimental design. In Experiment 1, Filters 1 and 2 were hybridized with ^{33}P-labeled, random hexamer generated cDNA fragments complementary to each of three RNA preparations (IH-G2490 RNA1–3) obtained from the cells of three individual cultures of strain IH-G2490 (Lrp$^+$). These three ^{33}P-labeled cDNA preparations were pooled prior to the hybridizations. Following phosphorimager analysis, these filters were stripped and hybridized with pooled, ^{33}P-labeled cDNA fragments complementary to each of three RNA preparations (IH-G2491 RNA1–3) obtained from strain IH-G2491 (Lrp$^-$). In Experiment 2, these same filters were again stripped and this protocol was repeated with ^{33}P-labeled cDNA fragments complementary to another set of three pooled RNA preparations obtained from strains IH-G2490 (IH-G2490 RNA 4–6) and IH-G2491 (IH-G2491 RNA 4–6) as described above. Another set of filters (Filter 3 and Filter 4) was used for Experiments 3 and 4 as described for Experiments 1 and 2. This protocol results in duplicate filter data for four experiments performed with the cDNA probes complementary to four independently prepared cDNA probe sets. Thus, since each filter contains duplicate spots for each ORF and duplicate filters are hybridized for each experiment, four measurements for each ORF are obtained from each of four experiments.

operon (ORFs in the same mRNA transcript) were detected, others were not. For example, hybridization with the message-specific targets detected signals above background for only three of the five genes of the *ilvGMEDA* operon, whereas hybridization with the random-hexamer-generated targets detected signals above background for all five genes of this operon. Furthermore, the expression level of the genes in each operon detected with the random-hexamer-generated targets usually varied less than threefold, while the expression levels of some of the genes of a common operon detected with the 3' message-specific primed targets either were not detected at all, or when they were detected they sometimes varied more than 1000-fold.

To explain why the message-specific-generated targets did not detect as many mRNAs as the random-hexamer-generated targets, we suggested that some of the cDNA targets derived from the message-specific primers did not hybridize to their cognate mRNA. This suggestion was based on the simple fact that it is difficult at best to design 4290 individual primer sequences that will hybridize to their cognate sequences with the same efficiency under a single hybridization condition. We also suggested that some of the primers might contain complementary sequences and hybridize to one another rather than the DNA array probes; and further, that some of these primers might contain internally complementary sequences and hybridize to themselves instead of the probes on the array.

To explain the observation that a wide variation of signals were observed with the message-specific-labeled targets for genes of a common operon, we suggested that a variable amount of ^{33}P was incorporated into each target because of unequal hybridization efficiencies and different lengths of labeled 3'-proximal cDNA fragments. On the other hand, since each mRNA (or mRNA fragment) is randomly primed with the random hexamers, the amount of ^{33}P label incorporated into each cDNA target should be largely proportional to the ORF length.

To test the validity of these suggestions, we reasoned that since the stoichiometry of each ORF in the genome is 1, and since each region of the chromosome is randomly primed with the random hexamers, ^{33}P-labeled cDNA targets prepared from genomic DNA with random hexamers should hybridize to all of the full-length ORF probes on the array and produce signal intensities largely proportional to the length of the ORF. At the same time, hybridization of the ^{33}P-labeled cDNA targets prepared with the 3'-specific primers should be dependent on a complex set of variables, including the length and the different hybridization efficiencies of each cDNA target generated with each ORF-specific primer.

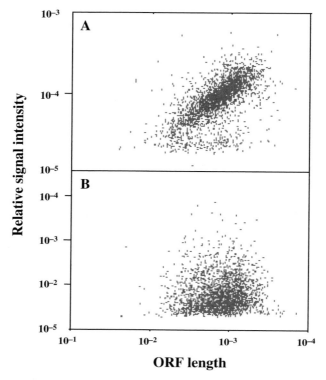

Figure 4.2. Relationships between the logarithm of hybridization signals and the logarithm of ORF lengths with targets prepared from genomic DNA. Scatter plots showing the relationships between hybridization signal intensities with ^{33}P-labeled cDNA probes generated from genomic DNA with random-hexamer oligonucleotides (A) or 3′-ORF-specific DNA primers (B).

The data presented in Figure 4.2 confirmed our expectations. DNA hybridization signals for each of the 4290 ORFs on the array are observed with the random-hexamer-labeled targets, while hybridization signals for only two-thirds of the ORFs on the array are observed with the message-specific primer labeled targets (the same ratio of ORF-specific versus random-hexamer labeled probe hybridization signals we observed with the cDNA targets generated from RNA). Thus, as we expected, one-third of the message-specific primers did not hybridize to their cognate genomic sequence, either because of hybridization conditions or because they hybridized to themselves or to one another. Also as expected, the data displayed in Figure 4.2 show that the hybridization signal for the random-hexamer labeled targets generated from genomic DNA is reasonably proportional to

ORF length ($r^2 = 0.41$). However, no significant correlation between ORF length and hybridization signal is observed with the ORF-specific labeled probes ($r^2 = 0.004$) (Figure 4.2).

Rapid turnover of mRNA in bacterial cells

An additional complication that contributes to the disparate results obtained with random hexamer and message-specific labeled targets pertains to the rapid turnover rates of mRNA in bacterial cells (from a few seconds to several minutes versus several hours to days in eukaryotes). In *E. coli* rapid mRNA decay is initiated by endonucleolytic cleavages followed by 3′ to 5′ exonucleolytic degradation; therefore, if the initial endonucleolytic site is adjacent to the 3′ ORF-specific primer binding site, this region is rapidly degraded and little or no steady-state message is extracted for primer extension labeling of this gene-specific transcript. On the other hand, if the 3′ ORF-specific primer binding site is located in a portion of the mRNA stabilized by secondary structure that is distant from the initial endonucleolytic site, it will be present in the cell at a high steady-state level (Figure 4.3). Therefore, since different parts of an mRNA molecule are degraded at widely differing rates, the 25 base pair region of each message complementary to each message-specific target can be present in the cell at different levels. Under these conditions varying amounts of message will be extracted for primer extension labeling of each gene-specific transcript depending on the location and degradation rate of the primer site. On the other hand, the random-hexamer-labeling procedure produces RNA–DNA duplexes for primer extension from all of the partial degradation products of each message. Since the exonucleolytic clearance of mRNA degradation products to free nucleotides follows endonucleolytic message inactivation (at a presumably constant enzymatic rate) the random hexamers detect the steady state level of all of these intermediate degradation products. This implies that although the functional half-lifes of *E. coli* mRNA are rapid and message specific, the "clearance" rate for message degradation intermediates to free nucleotides must occur at a more constant rate. Thus, the relative mRNA expression levels measured with the random hexamer labeled probes should be more closely related to their rates of synthesis and, therefore, their relative abundance in the cell. This conclusion is supported by two observations. First, nearly equal levels of expression of genes of a common operon are observed with the random-hexamer-generated targets, but not with the targets generated with the 3′ message-specific primers. Second, a positive correlation between mRNA and protein abundance is observed

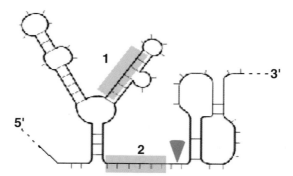

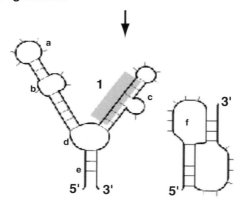

Figure 4.3. mRNA degradation in bacteria. The top structure represents the 3′ region of an mRNA molecule. The triangle represents an endonuclease enzyme positioned at its RNA cleavage site. The gray boxes denote two alternative complementary regions for binding of a 3′ message-specific primer oligonucleotide. Site 1 is protected from 3′ to 5′ exonucleolytic degradation following endonucleolytic cleavage because of its position in a secondary structure, whereas site 2 is readily susceptible to 3′ to 5′ exonucleolytic degradation. Higher *in vivo* steady-state levels of this mRNA would be detected with an oligonucleotide complementary to site 1 than to site 2.

with random-hexamer-generated targets, but this correlation is not observed when message-specific-generated targets are used [2].

As an aside, it might be thought that because DNA array hybridization data measured with random-hexamer-labeled probes shows a correlation with ORF length, each measurement should be corrected for this parameter. However, we do not feel that it is necessary to correct the expression data obtained with random-hexamer-labeled cDNA targets for probe ORF lengths. This is because the less than ten-fold variance in ORF lengths contributes less than 1 percent of the greater than four orders of magnitude of variance in expression level measurements obtained with the RNA-derived targets.

Preparation of bacterial targets for in situ *synthesized DNA arrays*

So, as it turns out, fears that random-hexamer-generated ^{33}P-labeled cDNA targets prepared from total RNA would generate unacceptable backgrounds are unwarranted, at least for pre-synthesized DNA arrays containing PCR-generated, full-length ORF probes. As an example, a nylon filter array manufactured by Sigma-Genosys containing duplicate spots of PCR-generated, full-length ORF probes of each of the 4290 genes of *E. coli* hybridized with ^{33}P-labeled cDNA targets synthesized with random hexamers from a total RNA preparation is shown in Figure 3.2. It can be seen that backgrounds are very low, especially when care is taken to remove traces of genomic DNA in the RNA preparations, a common source of background contamination. This contamination is particularly serious since genomic DNA, of course, contains sequences complementary to all of the ORFs. Thus, if care is not taken to remove all traces of genomic DNA it will hybridize to all of the ORFs and obscure the message-specific target signals. Another reason for the low background on pre-synthesized arrays pertains to the high temperature and stringency hybridization conditions that can be used with DNA arrays containing full length ORF probes. These stringent hybridization conditions (Appendix A) are possible because the average length of each probe on these arrays is about 850 nucleotides and sizing of the random-hexamer-generated targets shows that they are of an average length of 400 nucleotides [8]. The background is low because the stringency of these hybridization conditions is such that no non-mRNA labeled cDNA targets with probe sequence identities less than about 100 nucleotides can hybridize to the DNA array probes.

In spite of their advantages, the use of targets prepared from total RNA is not an option for gene expression profiling with *in situ* synthesized arrays such as Affymetrix GeneChips™ that contain short oligonucleotide (25 nucleotides) target-specific probes. In this case, much less stringent hybridization conditions must be used. Under these conditions, random-hexamer-directed target synthesis from a total RNA preparation produces labeled targets from abundant ribosomal and other non-mRNA targets that will cross-hybridize with all of the probes on the GeneChip™ and mask gene-specific measurements. Methods to circumvent this problem are discussed below.

Target preparation with non-polyadenylated mRNA from bacterial cells

Most *in situ* synthesized DNA arrays contain short oligonucleotide probes. For example, the *E. coli* Affymetrix GeneChip™ contains 15 25-mer oligonucleotide DNA probes perfectly complementary to sequences in each of 4241 ORFs. It also contains from 1 to 298 probes complementary to each of the 107 rRNAs, tRNAs, and small regulatory RNAs, as well as each of the 2886 intergenic ORF regions. Because of the short length of these probes, relatively low stringency hybridization conditions are required. Under these hybridization conditions, mismatch base pairing between perfect match (PM) probes and targets can occur that obscure experimental results. For this reason, each *E. coli* GeneChip™ also contains a 25-mer oligonucleotide mismatch (MM) probe differing from each corresponding PM probe by possessing a base pair mismatch in the middle of the probe. This allows an estimate of mismatch hybridization encountered by each probe. The hybridization signal for each target is measured as the average difference (AD) of the hybridization signals between each of the PM and MM probe pairs for each target (AD = Σ(PM − MM) / number of probe pairs). Since each mRNA species obtained from purified polyadenylated mRNA from eukaryotic cells is, by and large, unique, this method produces low background measurements for each target. However, bacterial mRNA is not polyadenylated, and mRNA represents less than 10 percent of the total RNA extracted from bacterial cells. Therefore, without separation of the mRNA from at least the bulk RNA (ribosomal RNA), the levels of mismatch hybridization of biotinylated RNA targets prepared from total RNA would be unacceptable.

To circumvent this problem, Affymetrix developed a procedure for enriching the mRNA to ribosomal RNA ratio of total RNA samples four-to fivefold (Figure 4.4). In this procedure, which we will see later is applicable to eukaryotic as well as prokaryotic total RNA samples, oligonucleotide primers specific for ribosomal RNA sequences are used to generate rRNA/cDNA hybrids. Next, the RNA moiety of these double stranded hybrids is digested away with RNase H. Finally, the cDNA strand is removed with DNase I (see Appendix A). Although high levels of hybridization to mismatch probes are observed, acceptable backgrounds and average difference signals between perfect match and mismatch probes are

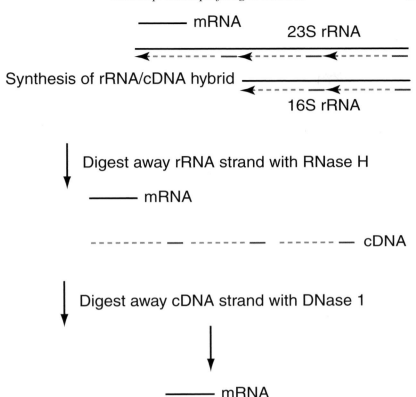

Figure 4.4. A poly(A) independent method for mRNA enrichment. Oligonucleotide primers (short blue lines) specific for ribosomal RNA sequences are used to generate rRNA/cDNA (represented by the red-dashed arrows) hybrids. The RNA moiety (black line) of these double-stranded hybrids is digested away with RNase H. Finally, the cDNA strand is removed with DNase I.

obtained with the Affymetrix *E. coli* GeneChip™ when these message-enriched RNA samples are used to generate biotin-labeled RNA targets (see Chapter 7). Typically, this procedure gets rid of about 90 percent of the ribosomal RNA in a bacterial total RNA preparation. Other protocols for the preparation and hybridization of prokaryotic and eukaryotic labeled targets to Affymetrix GeneChips™ as well as nylon filters and glass slide arrays also are described in Appendix A.

Problems associated with target preparation with polyadenylated mRNA from eukaryotic cells

To date, all reported eukaryotic DNA array experiments have been performed with targets prepared from poly(A)$^+$ mRNA isolated with oligo(dT) chromatography columns or by synthesizing cDNA targets with oligo(dT) primers. This has been done because, like the bacterial case, it has been feared that targets prepared from total RNA would generate unacceptable backgrounds. However, again like the bacterial case, ^{33}P- or fluorophore-labeled, random-hexamer-primed cDNA targets prepared from eukaryotic total mRNA preparations do not produce high backgrounds when they are hybridized under stringent conditions to DNA arrays containing full length ORF probes. In fact, several reports demonstrate that using poly(A)$^+$ RNA for eukaryotic gene expression profiling experiments leads to erroneous conclusions. The reason for this is that efficient oligo(dT) purification requires poly(A) tails at least 20 nucleotides in length. This is a problem because, at least in yeast, it has been demonstrated that during balanced growth, as much as 25–50 percent of the total mRNA exists with a poly(A) tail length less than 20 nucleotides. This mRNA, which is not efficiently recovered from oligo(dT) columns or primer-extended with oligo(dT) primers but can be detected in northern gels with message-specific probes, is classified as poly(A)$^-$ RNA. In other words, yeast mRNAs exist as a population with as much as 25–50 percent of the total mRNA pool in a poly(A)$^-$ form [9, 10]. Clearly, this confounds the results of any yeast gene expression profiling experiment performed with targets prepared from poly(A)$^+$ mRNA by oligo (dT) purification or priming.

This conclusion is supported in a recent report by Olivas and Parker [10]. They have demonstrated large discrepancies between the results of gene expression profiling experiments performed with the yeast *Saccharomyces cerevisiae* that compared targets prepared from purified poly(A)$^+$ and total RNA preparations. For example, they demonstrated that array experiments using targets prepared from poly(A)$^+$ mRNA imply that the expression level of the COX17 gene decreases more than tenfold in a strain containing a *puf3* deletion; however, northern analysis of total RNA with a COX17-specific probe show that this mRNA actually increases more than twofold in this strain. This over 20-fold discrepancy is explained by the fact that in the *puf3* deletion strain COX17 transcripts are partially deadenylated and only poorly selected by purification with oligo(dT). At the same time, this message is stabilized in the *puf3* deletion strain. The result is that while the COX17 message accumulates in the deletion strain it is not efficiently recovered by oligo(dT) purification and therefore appears to have decreased in the array experiment. Olivas and Parker [10] suggest that message-specific degradation mechanisms that affect polyadenylation levels are a major reason for discrepancies between DNA array measurements and subsequent northern blot results employing message-specific probes. In other words, if polyadenylation levels are differentially affected by different treatment conditions, then erroneous conclusions concerning gene expression levels from DNA array experiments performed with poly(A) derived targets might be reached. This suggests that gene expression profiling experiments in yeast should be performed with targets prepared from total RNA preparations.

This same problem exists for gene expression profiling in mammalian systems where it is also known that some mRNAs such as actin mRNA have poly(A) tails shorter than 20 nucleotides and mRNA stabilities are regulated by deadenylation mechanisms. However, unlike with yeast, no systematic experiments have been performed with mammalian systems to determine the fraction of such mRNAs in the total mRNA pool. Nevertheless, prudence suggests that targets should be prepared from total RNA rather than poly(A)$^+$ RNA whenever possible. The feasibility of this suggestion is supported by the results of Tan *et al.* [11]. They have demonstrated that glass slides arrayed with full-length PCR products of each of the 984 ORFs of *Chlamydia trachomatis* can be hybridized with random-hexamer-generated targets prepared from total RNA from infected mouse cells with low backgrounds. The targets for this experiment were random-hexamer-generated cDNA products of total RNA from *Chlamydia*-infected or uninfected mouse cells. Even though the *Chlamydia*-mRNA represented a small fraction of the total mouse cell RNA, and even though only a small fraction of the mouse cells were infected with *Chlamydia*, very little cross-hybridization of mouse-derived, random-hexamer-generated cDNA targets to the *Chlamydia*-specific probes was detected. This is because of the high stringency hybridization conditions that can be employed with arrays printed with full-length ORF PCR products as previously described for the *E. coli* system. Equally satisfying results have been obtained with random-hexamer-generated, ^{33}P-labeled cDNA targets prepared from total human RNA hybridized to custom, medium-density (9000 probes) DNA arrays containing full-length human ORF probes [12].

A further advantage of using total RNA for target preparation is that it minimizes experimental manipulations. This means that variations due to mRNA degradation and loss during the target preparation procedure are decreased. Also, even though mRNA half-lifes are much longer in mammalian than bacterial systems, any steady-state mRNA level differences that do arise because of differing message degradation rates in the 3′ portion of mammalian mRNAs will be minimized with labeled cDNA targets prepared from total RNA with random hexamer priming [2]. Furthermore, as with the bacterial system discussed above and in Chapter 7, target synthesis from total RNA preparations significantly improves quantitation of the measurements of the relative abundance of each mRNA species.

A total RNA solution for target preparation from eukaryotic cells

The most common method for the preparation of biotinylated RNA targets from poly(A)$^+$ RNA for eukaryotic gene expression profiling with Affymetrix GeneChips™ includes the following steps: (1) total RNA is extracted from cells or tissue; (2) poly(A)$^+$ mRNA is isolated with a cellulose oligo(dT) column; (3) cDNA is synthesized with oligo(dT) primers containing a T7 bacteriophage promoter; (4) the cDNA is used as a template for T7 RNA polymerase to synthesize RNA complementary to the genomic sense strand; (5) this cRNA is fragmented into molecules with an average length of 80–100 nucleotides; and (6) these cRNA molecules are biotinylated and hybridized to GeneChips™ containing sense strand probes. A common modification of this procedure is to skip the poly(A)$^+$ purification step and to proceed directly to the oligo(dT)-directed primer extension step. In either case, the problem of missing transcripts with short poly(A) tails described by Olivas and Parker [10] remains a serious issue. The solution of this problem offered above for pre-synthesized arrays containing full-length ORF probes, that is to prepare targets from total RNA preparations, is not applicable for *in situ* synthesized arrays containing short oligonucleotide probes. Under the low-stringency hybridization conditions required for these arrays, the levels of mismatch hybridization of biotinylated RNA targets prepared from cDNA generated from total RNA would be unacceptable.

One solution to isolate all of the mRNA in both the poly(A)$^+$ and the poly(A)$^-$ fractions while avoiding the oligo (dT)-dependent poly(A) mRNA purification and amplification steps would be to adopt the methods used for preparing targets from bacterial cells (Appendix A). In this procedure, ribosomal RNA-specific probes would be used to generate rRNA/cDNA hybrid duplexes. The rRNA moiety would be removed with RNase H and the rRNA-specific cDNA would be removed with DNase I. However, since the current Affymetrix GeneChips™ contain sense strand probes, it would not be possible to simply use hexamer-generated sense strand cDNA targets prepared from this mRNA enriched preparation for hybridization to these arrays. This requires an extra step. That is to use random oligonucleotides containing a T7 bacteriophage promoter to generate antisense cRNA from the cDNA generated in the previous step. This cRNA could then be fragmented, biotin labeled, and hybridized to the currently available sense strand probe arrays. Protocols similar to those described for the bacterial system for the preparation of biotin-labeled cRNA targets as suggested here are described in Appendix A. Methods for the preparation of ^{33}P- or fluorophore-labeled cDNA targets from total eukaryotic RNA using random hexamers for hybridization to nylon filter or glass slide DNA arrays containing full length ORF probes also are described in Appendix A.

Target cDNA synthesis and radioactive labeling for pre-synthesized DNA arrays

Although radioisotope-labeled targets can be measured with greater sensitivity and over a greater linear range than fluorophore-labeled targets, their use has been limited to large-format arrays. There are two primary technical reasons for this. First, the 50 μm limitation in the scanning resolution of most commercially available phosphorimagers (described in Chapter 3) requires probe sites with diameters of at least 250 μm for reproducible signal quantitation. Second, because the omni-directional radioactive emissions from targets hybridized to one spot on an array can contaminate the signals obtained from neighboring probe sites, array formats containing probe sites spaced closer than about 300 μm are not feasible. Nevertheless, target labeling with radioisotopes is the method of choice for large-format arrays such as nylon filters where probe features as large as 1 mm in diameter are separated from one another by as much as 2 mm.

Data acquisition for nylon filter experiments

Once a gene expression profiling experiment has been completed, methods for extracting the data from each feature on the array must be employed. For nylon-filter-based experiments with ^{33}P-labeled target values, the filter is exposed to a phosphorimaging screen and the target intensities are visualized by scanning in a phosphorimager that generates a 16-bit TIFF (or proprietary format) image (Figure 3.2). At this point, software to extract and digitize the pixel density of each array feature in the image file and to correlate each feature with the correct ORF is required. Several commercial software packages are available for these purposes (Table 4.1). In the

Table 4.1. *Commercial sources for DNA array data acquisition software*

Name	Company	Web address	Type of processed image	Fluorescent/ radioactive	Type of array	Spot address correlated to gene name?
QuantArray for Windows NT	Packard Bioscience	www.packardbiochip.com	TIFF Bitmap	Fluorescent	Glass slide	Yes
Microarray Suite for Mac/Win	Scanalytics	www.scanalytics.com	TIFF	Both	Glass slide, chip, filter	Yes
ArrayVision v 4.0 Win NT	Imaging Reseach Inc.	www.imagingresearch.com	TIFF	Both	Filter, glass slide	Yes
AtlasImage v 2.01 Win	Clontech Inc.	www.clontech.com	TIFF GEL	Both	Filter, glass slide	Yes
GeneTAC Integrator Software, Win	Genomic Solutions Inc.	www.genomicsolutions.com	TIFF	Fluorescent	Glass slide	Yes
ImaGene v4.1 Win/Mac(beta)	BioDiscovery Inc.	www.biodiscovery.com	TIFF GEL BAS	Both	Glass slide, filter, membrane	Yes

Software	Company	URL	File format	Label	Array	Analysis
ArrayPro Analyzer, Win	Media Cybernetics	www.mediacy.com	TIFF	Fluorescent	Glass slide	No
GenePix Pro 3.0.5, Win	Axon	www.axon.com	TIFF	Fluorescent	Glass slide	Yes
ScanAlyze Win/95/98/ NT 4.0	Eisen Lab/ Stanford University	rana.lbl.gov/EisenSoftware.htm	TIFF	Fluorescent	Glass slide	No
Microarray Suite Win NT 4.0	Affymetrix	www.affymetrix.com	TIFF	Fluorescent	Affymetrix chip	Yes
Pathways™ Win/Linux/Unix	ResGen	www.resgen.com	TIFF GEL	Radioactive	Filter	Yes
ArrayEase Win	AlphaInnotech	www.alphainnotech.com	TIFF	Fluorescent	Glass slide	Yes

experiments described in Chapter 7, we used DNA ArrayVision software, obtained from Imaging Research Inc. (St. Catharines, Ontario, Canada) to provide software to prepare an array template, to record the pixel density, and to perform the background subtraction for each of the 18 432 addresses on Sigma-Genosys Panorama™ *E. coli* filters. A total of 8580 of the features on these filters contain duplicate copies of a full-length PCR product of each of the 4290 *E. coli* ORFs. We used the remaining interspersed 9852 empty addresses for background measurements. The backgrounds are usually quite low and constant over the surface of the filter. In this case, a global average background measurement is subtracted from each experimental measurement. If the background varies in certain areas of the filter, the software can calculate and apply local background corrections.

It is important to be aware of the fact that some array analysis software packages only correlate target intensities values with filter addresses, not with gene names. This laborious task can either be performed in Excel with worksheets supplied by the array manufacturer, or by custom software programs such as the ArrayVision Data Conversion Interface program available for online use at the Functional Genomics internet site of the University of California, Irvine.

Data acquisition for Affymetrix GeneChip™ glass slide experiments

Because of the high density of probe features on Affymetrix GeneChips™ and glass slide arrays, confocal laser scanners described in Chapter 3 are required to accurately measure target signal intensities. Since these scanners operate at resolutions as high as 3 μm, probe feature diameters as small as 15 μm can be accurately measured. As an example, the typical feature size on current Affymetrix GeneChips™ is 24×24 μm. In this case, each GeneChip™ array is scanned twice with an HP GeneArray™ confocal laser or equivalent scanner, typically at a 3 μm resolution, and the intensities at each probe site from both scans are averaged and saved in a *.CEL file. In addition, an array image file is created and saved with a *.DAT extension. The data in the *.CEL file is used to determine the background subtracted average difference of each probe set $(AD = \Sigma(PM - MM)$ / number of probe pairs) as described earlier. These raw, background-subtracted average difference values for each probe set are used as an indicator of the level of expression of each gene for statistical analysis with an appropriate software package such as Microsuite provided by Affymetrix or the CyberT statistical program discussed in Chapters 5 and 7 available for online use at www.genomics.uci.edu.

Normalization methods

Before we can determine the differential gene expression profiles between two conditions obtained from the data of two DNA array experiments, we must first ascertain that the data sets are comparable. That is, we must develop methods to normalize data sets in a way that accounts for sources of experimental and biological variations, such as those discussed above, that might obscure the underlying variation in gene expression levels attributable to biological effects. However, with few exceptions, the sources of these variations have not been measured and characterized. As a consequence, many array studies are reported without statistical definitions of their significance. This problem is even further exacerbated by the presence of many different array formats and experimental designs and methods. While some theoretical studies that address this important issue have appeared in the literature, the normalization methods currently in common use are based on more pragmatic biological considerations. Here we consider the pros and cons of these data normalization methods and their applicability to different experimental designs and DNA array formats. Basically, these methods attempt to correct for the following variables:

- number of cells in the sample
- total RNA isolation efficiency
- mRNA isolation and labeling efficiency
- hybridization efficiency
- signal measurement sensitivity.

Normalization to total or ribosomal RNA

In some reports, the authors assume the total RNA in a cell remains constant under all conditions. This is based on the belief that ribosomal RNA is constituitively expressed in all cells. Thus, since it comprises more than 90 percent of the total RNA they conclude that it can be used as a normalization standard for small fluctuations in mRNA levels. In this case, the data from independent experiments can be normalized to one another by adjusting their gene expression levels to the amount of total or ribosomal RNA harvested from each sample. The problem with this method is that, as recently pointed out by Suzuki *et al.* [13], the underlying biological assumption is wrong. Different cell types or cells growing in different conditions can produce widely differing amounts of total RNA. Indeed it is well known that total RNA production in mammalian cells can vary by as much as 50-fold.

Normalization to housekeeping genes

This method is based on the assumption that certain "housekeeping" genes such as GAPDH or actin are not regulated in growing cells. In spite of numerous reports that under certain circumstances this is not the case [13], this remains a popular method that can be appropriate for some experimental designs. For example, if one is comparing the expression levels of a small number of genes where other more global normalization methods are inappropriate this method is acceptable. However, even in these cases, it should be kept in mind that small differences could be attributable to differences in the standard. In general, this should not be considered an acceptable method for the normalization of DNA array data.

Normalization to a reference RNA

A widely used method for all types of DNA arrays and experimental designs is to add a given amount of an RNA standard from another organism to each total RNA sample. For example, Affymetrix adds bacterial and bacteriophage probes to their eukaryotic GeneChips™. This allows the data from each array to be scaled to the levels of these known amounts of RNA standards. The major advantage of this method is that by comparison to the standards it allows an estimate of the absolute level of targets in each experimental sample. Another advantage of this method is that it effectively normalizes for variances in mRNA isolation, labeling, and hybridization efficiencies. The disadvantages arise from the compounded errors associated with the amounts of the standard applied to each array and errors in the measurements of these standards.

Normalization by global scaling

The most widely used data normalization method that relies on the fewest assumptions and addresses all of the variables listed above is the global normalization method. Here, the sum of the measurements of each array are simply scaled to a common number. For example in the experiments described in Chapter 7 we divide each measurement from each DNA array by the sum of all measurements of that array. This scales the total signal on all arrays to a value of one with the advantage that each individual measurement is expressed as a fraction of the total signal, in other words as the fraction of total mRNA. While this is a simple method that performs well, it also has its drawbacks. It assumes that the total amount of mRNA remains constant under various experimental conditions. If one is examining the gene expression profile of an entire genome this is probably a safe assumption, especially for bacterial cells. In this case, the large majority of genes remain relatively constant and those genes whose expression levels are increased are countered by genes whose expression levels decrease. However, this is a tenuous assumption at best if a small subset of functionally related genes or if specialized cell types are queried.

Since none of these methods accommodates the many sources of variables encountered in DNA array experiments, this is an intense area of bioinformatics research. For example, Li and Wong [14] have considered the variance introduced by the use of different probes to interrogate the same gene in Affymetrix GeneChips™. They have developed model-based algorithms for identifying and handling cross-hybridizing probes that greatly improve the interpretation of GeneChip™ data, especially when only a few replicate experiments are performed (see Chapter 7). As DNA array formats become more standardized as more research in this area is performed it can be anticipated that general solutions for these basic problems will emerge.

REFERENCES

1. DeRisi, J. L., Iyer, V. R., and Brown, P. O. The metabolic and genetic control of gene expression on a genomic scale. 1997. *Science* 278(5338):680–686.
2. Arfin, S. M., Long, A. D., Ito, E., Riehle, M. M., Paegle, E. S., and Hatfield, G. W. Global gene expression profiling in *Escherichia coli* K12: the effects of integration host factor. 2000. *Journal of Biological Chemistry* 275:29672–29684.
3. Schena, M., Shalon, D., Davis, R. W., and Brown, P. O. Quantitative monitoring of gene expression patterns with a complementary DNA microarray. 1995. *Science* 270:467–470.
4. Schena, M. Genome analysis with gene expression microarrays. Review. 1996. *Bioessays* 18:427–431.
5. Rhee, K., Parekh, B. S., and Hatfield, G. W. Leucine-responsive regulatory protein–DNA interactions in the leader region of the ilvGMEDA operon of *Escherichia coli*. 1996. *Journal of Biological Chemistry* 271:26499–26507.
6. Dolter, K. E., and Braman, J. C., Small-sample total RNA purification: laser capture microdissection and cultured cell applications. 2001. *Biotechniques* 6:1358–1361.
7. Neidhardt, F. C., Bloch, P. L., and Smith, D. F. Culture medium for enterobacteria. 1974. *Journal of Bacteriology* 119:736–747.
8. Ito, E. T., and Hatfield, G. W., unpublished data.
9. Decker, C. J., and Parker, R. Mechanisms of mRNA degradation in eukaryotes. Review. 1994. *Trends in Biochemical Sciences* 19:336–340.
10. Olivas, W., and Parker, R. The Puf3 protein is a transcript-specific regulator of mRNA degradation in yeast. 2000. *EMBO Journal* 19:6602–6611.
11. Yu, H., Tan, M., and Hatfield, G.W., unpublished data.
12. Vawter, M., and Hatfield, G. W., unpublished data.

13. Suzuki, T., Higgins, P. J., and Crawford, D. R. Control selection for RNA quantitation. 2000. *Biotechniques* 29:332–337.
14. Li, C., and Wong, W. H. Model-based analysis of oligonucleotide arrays: expression index computation and outlier detection. 2001. *Proceedings of the National Academy of Sciences of the USA* 98:31–36.

5

Statistical analysis of array data: Inferring changes

Problems and common approaches

Although many data analysis techniques have been applied to DNA array data, the field is still evolving and the methods have not yet reached a level of maturity [1]. Even very basic issues of signal-to-noise ratios are still being sorted out.

Gene expression array data can be analyzed on at least three levels of increasing complexity. The first level is that of single genes, where one seeks to establish whether each gene in isolation behaves differently in a control versus an experimental or treatment situation. Here experimental/treatment is to be taken, of course, in a very broad sense: essentially any situation different from the control. Differential single-gene expression analysis can be used, for instance, to establish gene targets for drug development. The second level is multiple genes, where clusters of genes are analyzed in terms of common functionalities, interactions, co-regulation, etc. Gene co-expression can provide, for instance, a simple means of gaining leads to the functions of many genes for which information is not available currently. This level includes also leveraging DNA array data information to analyze DNA regulatory regions and finding regulatory motifs. Finally, the third level attempts to infer and understand the underlying gene and protein networks that ultimately are responsible for the patterns observed. Other issues of calibration, quality control, and comparison across different experiments and technologies are addressed in Chapter 7 (see, for instance, also [2, 3]).

It should be clear that this classification is useful only as an initial roadmap, and there are other possible viewpoints from which array data can be analyzed, notwithstanding that the boundaries between levels and between problems are fuzzy. In particular, this classification takes a gene-centric viewpoint. Indeed, one could, for instance, look also at the data from the angle of the conditions and ask how similar are two conditions from a gene expression standpoint [4], or cluster data according to experiments. If for nothing else than convenience, we shall still follow the simple roadmap. Thus, in this chapter we deal with the first problem of inferring significant gene changes. The treatment here largely follows our published work [5, 6].

To begin with, we assume for simplicity that for each gene X the data D consists of a set of measurements $x_1^c,...,x_{nc}^c$ and $x_1^t,...,x_{nt}^t$ representing expression levels, or rather their logarithms, in both a control and treatment situation. For each gene, the fundamental question we wish to address is whether the level of expression is significantly different in the two situations.

One approach commonly used in the literature at least in the first wave of publications (see for instance, [7, 8, 9]), has been a simple-minded fold approach, in which a gene is declared to have significantly changed if its average expression level varies by more than a constant factor, typically two, between the treatment and control conditions. Inspection of gene expression data suggests, however, that such a simple "twofold rule" is unlikely to yield optimal results, since a factor of two can have quite different significance and meaning in different regions of the spectrum of expression levels, in particular at the very high and very low ends.

Another approach to the same question is the use of a t-test, for instance on the logarithm of the expression levels. This is similar to the fold approach because the difference between two logarithms is the logarithm of their ratio. This approach is not necessarily identical to the first because the logarithm of the mean is not equal to the mean of the logarithms; in fact it is always strictly greater by convexity of the logarithm function. The two approaches are equivalent only if one uses the geometric mean of the ratios rather than the arithmetic mean. In any case, with a reasonable degree of approximation, a test of the significance of the difference between the log expression levels of two genes is equivalent to a test of whether or not their fold change is significantly different from 1.

In a t-test, the empirical means m_c and m_t and variances s_c^2 and s_t^2 are used to compute a normalized distance between the two populations in the form:

$$t = (m_c - m_t)/\sqrt{\frac{s_c^2}{n_c} + \frac{s_t^2}{n_t}} \tag{5.1}$$

where, for each population, $m = \Sigma_i x_i/n$ and $s^2 = \Sigma_i (x_i - m)^2/(n-1)$ are the well-known estimates for the mean and standard deviation. It is well known in the statistics literature that t follows approximately a Student distribution (Appendix A), with

$$f = \frac{[(s_c^2/n_c) + (s_t^2/n_t)]^2}{\dfrac{(s_c^2/n_c)^2}{n_c - 1} + \dfrac{(s_t^2/n_t)^2}{n_t - 1}} \tag{5.2}$$

degrees of freedom. When t exceeds a certain threshold depending on the confidence level selected, the two populations are considered to be different. Because in the t-test the distance between the population means is normalized by the empirical standard deviations, this has the potential for addressing some of the shortcomings of the simple fixed fold-threshold approach. The fundamental problem with the t-test for array data, however, is that the repetition numbers n_c and/or n_t are often small because experiments remain costly or tedious to repeat, even with current technology. Small populations of size $n = 1$, 2 or 3 are still very common and lead, for instance, to poor estimates of the variance. Thus a better framework is needed to address these shortcomings.

Here we describe a Bayesian probabilistic framework for array data, which bears some analogies with the framework used for sequence data [10] and can effectively address the problem of detecting gene differences.

Probabilistic modeling of array data

The Bayesian probabilistic framework

Several decades of research in sequence analysis and other areas have demonstrated the advantages and effectiveness of probabilistic approaches to biological data. Clearly, current DNA array data is inherently very noisy, due to experimental and biological variables discussed in Chapter 4 that are difficult to control. But in addition, biological data sets are characterized by high degrees of variability that go well beyond the level of measurement noise. In sequence data, for instance, measurement noise levels can be reduced practically to zero in most cases (highly repetitive regions can be an exception) by sequencing the same region multiple times. Nevertheless, probabilistic models of sequence data, such as graphical models and Markov models [10, 11] increasingly play a fundamental role. This is because variability is an integral part of biological systems – brought about by, among others, eons of evolutionary tinkering. Even if measurements were close to perfect, there is some degree of stochasticity in the gene-regulation machinery itself.

Biological systems also have very high dimensionality: even in a large array experiment, only a very small subset of relevant variables is measured, or even under control. The vast majority of variables remains hidden and must be inferred or integrated out by probabilistic methods. Thus array data requires a probabilistic approach because it is highly noisy and variable, and many relevant variables remain unobserved behind the massive data sets.

This modeling problem is properly addressed using a Bayesian probabilistic framework. The general Bayesian statistical framework codifies how to proceed with data analysis and inference in a rational way in the presence of uncertainty. Under a small set of common sense axioms, it can be shown that subjective degrees of belief must obey the rules of probability and proper induction must proceed in a unique way, by propagation of information through Bayes' theorem. In particular, at any given time, any hypothesis or model M can be assessed by computing its posterior probability in light of the data according to Bayes' formula: $P(M|D) = P(D|M)P(M)/P(D)$, where $P(D|M)$ is the data likelihood and $P(M)$ is the prior probability capturing any background information one may have. The fundamental importance of Bayes' theorem is that it allows inversion, going from the probability a given hypothesis assigns to the data (likelihood) to the probability of the hypothesis itself, given the data (posterior). Increasingly, the Bayesian framework is being applied successfully to a variety of data-rich domains. Whether one subscribes or not to the axioms and practices of Bayesian statistics [12, 13, 14, 15], it is wise to model biological data in general, and array data in particular, in a probabilistic way for the reasons outlined above.

In order to develop a probabilistic approach for array data, the lessons learnt with sequence data are worth remembering. In sequence data, the simplest probabilistic model is a first-order Markov model, or a die, associated with the average composition of the family of DNA, RNA, or protein sequences under study (Figure 5.1). The next level of modeling complexity is a Markov model with one die per position or per column in a multiple alignment. The die model is the starting point of all more sophisticated models used today, such as higher-order Markov models, hidden Markov models, and so forth. In spite of their simplicity, these die models are still useful as a background against which the performances of more sophisticated models are assessed.

Gaussian model for array data

In array data, the simplest model would assume that all data points are independent of each other and extracted from a single continuous distribu-

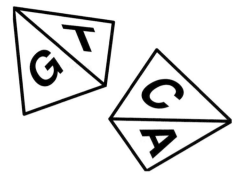

Figure 5.1. DNA dice.

tion, for instance a Gaussian distribution. While trivial, this "Gaussian die" model still requires the computation of interesting quantities, such as the average level of activity and its standard deviation, which can be useful to calibrate or assess global properties of the data. The next equivalent level of modeling is a set of independent distributions, one for each dimension, i.e., for instance each gene. While it is obvious that genes interact with each other in complex ways and therefore are not independent, the independence approximation is still useful and underlies *any* attempt, probabilistic or other, to determine whether expression level differences are significant on a *gene-by-gene* basis.

Here we first assume that the expression levels of a gene measured multiple times under the same experimental conditions will have a roughly Gaussian distribution. In our experience, with common technologies this assumption is reasonable, especially for the *logarithm* of the expression levels, corresponding to lognormal raw expression levels. To the best of our knowledge, large-scale replicate experiments have not been carried out yet to make very precise assessments. But at least to a first degree of approximation the normality assumption is correct. It is clear, however, that other distributions, such as gammas or mixtures of Gaussians/gammas, could be introduced at this stage. These would impact the details of the analysis (see also [16, 17]), but not the general Bayesian probabilistic framework.

Thus, in what follows we assume that the data has been pre-processed – including taking logarithms if needed – to the point where we can model the corresponding measurements of each gene in each situation (treatment or control) with a normal distribution $\mathcal{N}(x; \mu, \sigma^2)$. For each gene and each condition, we have a two-parameter model $w = (\mu, \sigma^2)$, and by focusing on one such model we can omit indices identifying the gene or the condition.

Assuming that the observations are independent, the likelihood is given by:

$$P(D|\mu, \sigma^2) \approx \prod_i \mathcal{N}(x_i; \mu, \sigma^2)$$

$$= C(\sigma^2)^{-n/2} e^{-\sum_i (x_i - \mu)^2/2\sigma^2}$$

$$= C(\sigma^2)^{-n/2} e^{-(n(m-\mu)^2 + (n-1)s^2)/2\sigma^2} \qquad (5.3)$$

where i ranges over replicate measurements. In this chapter, we write C to denote the normalizing constant of any distribution. The likelihood depends only on the sufficient statistics n, m, and s^2. In other words, all the information about the sample that is relevant for the likelihood is summarized in these three numbers. The case in which either the mean or the variance of the Gaussian model is supposed to be known is of course easier and is well studied in the literature [12, 13, 14].

A full Bayesian treatment requires introducing a prior $P(\mu, \sigma^2)$. The choice of a prior is part of the modeling process, and several alternatives [12, 13, 14] are possible, a sign of the flexibility of the Bayesian approach rather than its arbitrariness. Several kinds of priors for the mean and variance of a normal distribution have been studied in the literature [12, 13, 14], including non-informative improper prior, semi-conjugate prior, and conjugate prior distributions. A prior is said to be conjugate if it has the same functional form as the corresponding posterior. In our experience, so far the most suitable and flexible prior for array data is the conjugate prior, which adequately captures several properties of DNA array data including the fact that μ and σ^2 are generally *not* independent.

The conjugate prior

When estimating the mean alone of a normal model of known variance, the obvious conjugate prior is also a normal distribution. When estimating the standard deviation alone of a normal model of known mean, the conjugate prior is a scaled inverse gamma distribution (equivalent to $1/\sigma^2$ having a gamma distribution; see Appendix A). Intuitively, if you are not familiar with gamma distributions, think of this prior as a bell-shaped distribution, restricted to positive values only, on the variance. The form of the likelihood in Equation 5.3 shows that when both the mean and variance are unknown, the conjugate prior density must also have the form $P(\mu|\sigma^2)P(\sigma^2)$, where the marginal $P(\sigma^2)$ has a scaled inverse gamma distribution and the conditional distribution $P(\mu|\sigma^2)$ is normal.

This leads to a hierarchical model with a vector of four hyperparameters for the prior $\alpha = (\mu_0, \lambda_0, \nu_0 \text{ and } \sigma_0^2)$ with the densities:

$$P(\mu|\sigma^2) = \mathcal{N}(\mu; \mu_0, \sigma^2/\lambda_0) \tag{5.4}$$

and

$$P(\sigma^2) = I(\sigma^2; \nu_0, \sigma_0^2) \tag{5.5}$$

The expression for the prior $P(\mu, \sigma^2) = P(\mu, \sigma^2|\alpha)$ is then

$$C\sigma^{-1}(\sigma^2)^{-(\nu_0/2+1)} \exp\left[-\frac{\nu_0}{2\sigma^2}\sigma_0^2 - \frac{\lambda_0}{2\sigma^2}(\mu_0 - \mu)^2\right] \tag{5.6}$$

The expectation of the prior is finite if and only if $\nu_0 > 2$. Notice that it makes perfect sense with array data to assume a priori that μ and σ^2 are *dependent*, as suggested immediately by visual inspection of typical array data sets (see Figure 5.3, below). The hyperparameters μ_0 and σ^2/λ_0 can be interpreted as the location and scale of μ, and the hyperparameters ν_0 and σ_0^2 as the degrees of freedom and scale of σ^2. After some algebra, the posterior has the same functional form as the prior

$$P(\mu, \sigma^2|D, \alpha) = \mathcal{N}(\mu; \mu_n, \sigma^2/\lambda_n)I(\sigma^2; \nu_n, \sigma_n^2) \tag{5.7}$$

with

$$\mu_n = \frac{\lambda_0}{\lambda_0 + n}\mu_0 + \frac{n}{\lambda_0 + n}m \tag{5.8}$$

$$\lambda_n = \lambda_0 + n \tag{5.9}$$

$$\nu_n = \nu_0 + n \tag{5.10}$$

$$\nu_n\sigma_n^2 = \nu_0\sigma_0^2 + (n-1)s^2 + \frac{\lambda_0 n}{\lambda_0 + n}(m - \mu_0)^2 \tag{5.11}$$

The parameters of the posterior combine information from the prior and the data in a sensible way. The mean μ_n is a convex weighted average of the prior mean and the sample mean. The posterior degree of freedom ν_n is the prior degree of freedom ν_0 plus the sample size n, and similarly for the scaling factor λ_n. The posterior sum of squares $\nu_n\sigma_n^2$ is the sum of of the prior sum of squares $\nu_0\sigma_0^2$, the sample sum of squares $(n-1)s^2$, and the residual uncertainty provided by the discrepancy between the prior mean and the sample mean.

While is is possible to use a prior mean μ_0 for gene expression data, in many situations it is sufficient to use $\mu_0 = m$. The posterior sum of squares is then obtained precisely as if one had ν_0 additional observations all associated with deviation σ_0^2. While superficially this may seem like setting the prior after having observed the data, it can easily be justified [18]. Furthermore, a similar effect is obtained using a preset value μ_0 with $\lambda_0 \to 0$, i.e, with a very broad standard deviation so that the prior belief about the location of the mean is essentially uniform and vanishingly small. The selection of the hyperparameters for the prior is discussed in more detail below.

It is not difficult to check that the conditional posterior distribution of the mean $P(\mu|\sigma, D, \alpha)$ is normal $\mathcal{N}(\mu_n, \sigma^2/\lambda_n)$. The marginal posterior $P(\mu|D, \alpha)$ of the mean is Student $t(\nu_n, \mu_n, \sigma_n^2/\lambda_n)$ and the marginal posterior $P(\sigma^2|D, \alpha)$ for the variance is scaled inverse gamma $I(\nu_n, \sigma_n^2)$.

Finally, it is worth remarking that, if needed, more complex priors could be constructed using mixtures of conjugate priors, leading to mixtures of conjugate posteriors.

Full-Bayesian treatment versus hypothesis testing

The posterior distribution $P(\mu, \sigma^2|D, \alpha)$ is the fundamental object of Bayesian analysis and contains the relevant information about *all* possible values of μ and σ^2. At this stage of modeling, each gene is associated with two models $w_c = (\mu_c, \sigma_c^2)$ and $w_t = (\mu_t, \sigma_t^2)$; two sets of hyperparameters α_c and α_t; and two posterior distributions $P(w_c|D, \alpha_c)$ and $P(w_t|D, \alpha_t)$ (Figure 5.2). A full probabilistic treatment would require introducing prior distributions over the hyperparameters. These could be integrated out to obtain the true posterior probabilities $P(w_c|D)$ and $P(w_t|D)$.

The posterior distributions contain much more information than simple parameter point estimates, or the results of a simple *t*-test. The Bayesian approach could in principle detect interesting changes that are beyond the scope of the *t*-test. For instance, a gene with the same mean but a very different variance between the control and treatment situations goes undetected by a *t*-test, although the change in variance might be biologically relevant and correspond to a gene with oscillatory expression levels, with different amplitudes around the same mean (Figure 5.2). These and other more subtle differences could be captured by measuring the difference, for instance in terms of relative entropy [10], between the two posterior distributions.

The very idea of deciding whether a gene behaves differently between different conditions is somewhat at odds with the Bayesian framework which assigns a richer posterior distribution to all possible parameter values and in the case of continuous distributions, a typical mass of 0 to a single point event. Unless additional decision criteria are provided, a Bayesian treatment of the non-Bayesian decision problem requires

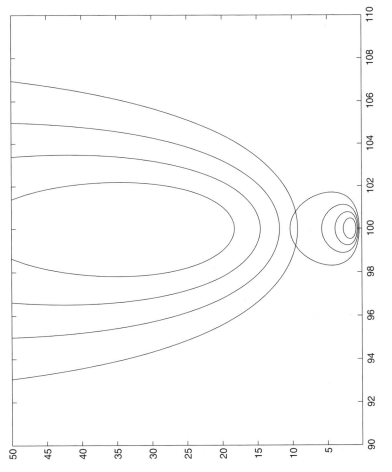

Figure 5.2. Contour plots for the posterior distribution of two hypothetical genes, or the same gene in two different conditions, using the posterior defined above seen in the $x - y$ plane, with $x = \mu$ and $y = \sigma^2$. In both cases, the mean is equal to 100 therefore the two situations are indistinguishable with a t-test or a fold approach. Yet the behaviors are very different in terms of standard deviations.

assigning a non-zero prior to the null hypothesis, which seems quite contrived (see section "Hypothesis testing" in Appendix B).

To address this decision issue, here we use a compromise between hypothesis testing and the more general Bayesian framework by leveraging the simplicity of the t-test, but using parameter and hyperparameter point estimates derived from the Bayesian framework.

Parameter point estimates

In order to perform the t-test we need to collapse the information-rich posterior distribution into single point estimates of the mean and variance of the expression level of a gene in a given situation. This can be done in a number of ways. In general, the most robust answer is obtained using the mean of the posterior (MP) estimate. An alternative is to use the mode of the posterior, or MAP (maximum a posteriori) estimate. For completeness, we derive both kinds of estimates.

By integration, the MP estimate is given by

$$\mu = \mu_n \quad \text{and} \quad \sigma^2 = \frac{\nu_n}{\nu_n - 2}\sigma_n^2 \tag{5.12}$$

provided $\nu_n > 2$. If we take $\mu_0 = m$, we then get the following MP estimate:

$$\mu = m \quad \text{and} \quad \sigma^2 = \frac{\nu_n \sigma_n^2}{\nu_n - 2} = \frac{\nu_0 \sigma_0^2 + (n-1)s^2}{\nu_0 + n - 2} \tag{5.13}$$

provided $\nu_0 + n > 2$. This is the default estimate implemented in the CyberT software described below. From Equation 5.7, the MAP estimates are:

$$\mu = \mu_n \quad \text{and} \quad \sigma^2 = \frac{\nu_n \sigma_n^2}{\nu_n - 1} \tag{5.14}$$

If we use $\mu_0 = m$, these reduce to:

$$\mu = \mu_n = m \quad \text{and} \quad \sigma^2 = \frac{\nu_n \sigma_n^2}{\nu_n - 1} = \frac{\nu_0 \sigma_0^2 + (n-1)s^2}{\nu_0 + n - 1} \tag{5.15}$$

Here the modes of the marginal posterior are given by

$$\mu = \mu_n \quad \text{and} \quad \sigma^2 = \frac{\nu_n \sigma_n^2}{\nu_n + 2} \tag{5.16}$$

In practice, Equations 5.13 and 5.15 give similar results and can be used with gene expression arrays. The slight differences between the two closely matches what is seen with Dirichlet priors on sequence data [10], Equation 5.13 being generally a slightly better choice. The Dirichlet prior is equivalent to the introduction of pseudo-counts to avoid setting the probability of any unobserved amino acid or nucleotide to zero. In array data, few observation points are likely to result in a poor estimate of the variance. With a single observation ($n = 1$), for instance, it would be unreasonable to think that the corresponding variance is zero; hence the need for regularization, which is achieved by the conjugate prior. In the MP estimate, the empirical variance is modulated by ν_0 "pseudo-observations" associated with a background variance σ_0^2. In summary, we are now modeling the expression level of a gene by a normal distribution with mean equal to the empirical mean and variance given by Equation 5.13. This requires estimating the hyperparameters σ_0^2 and ν_0, representing the background variance and its strength.

Hyperparameter point estimates and implementation

There are several possibilities for dealing with hyperparameters in general and in specific cases [5, 18, 19]. Here we describe a solution implemented in the CyberT web server [9] accessible through: www.genomics.uci.edu (Appendix C). In this approach, we use the *t*-test with the regularized standard deviation of Equation 5.13 and the number of degrees of freedom associated with the corresponding augmented populations of points, which incidentally can be fractional.

In the simplest case, where we use $\mu_0 = m$, one must select the values of the background variance σ_0^2, and its strength ν_0. The value of ν_0 can be set by the user. The smaller n, the larger ν_0 ought to be be. A simple rule of thumb is to assume that $l > 2$ points are needed to estimate properly the standard deviation and keep $n + \nu_0 = l$. This allows for a flexible treatment of situations in which the number n of available data points varies from gene to gene. A reasonable default is to use $l = 10$. A special case can be made for genes with activity levels close to the minimal detection level of the technology being used. The measurements for these genes being particularly unreliable, it may be appropriate to use a stronger prior for them with a higher value of ν_0.

For σ_0, one could use the standard deviation of the entire set of observations or, depending on the available knowledge, of particular categories of "similar" genes. In a flexible implementation, the background standard deviation is estimated by pooling together all the neighboring genes contained in a window of size *ws*. CyberT automatically ranks the expression levels of all the genes and lets the user choose this window size. The default is $ws = 101$, corresponding to 50 genes immediately above and below the average expression level of the gene under consideration.

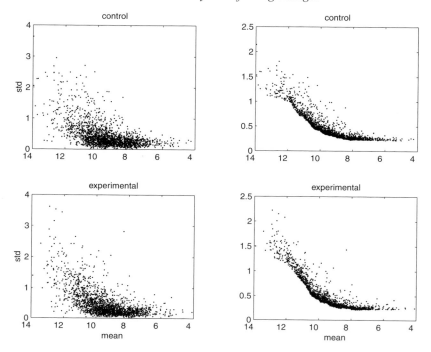

Figure 5.3. DNA array experiment on *Escherichia coli*. Data obtained from reverse transcribed ^{33}P-labeled RNA hydridized to commercially available nylon arrays (Sigma Genosys) containing each of the 4290 predicted *E. coli* genes. The sample included a wild-type strain (control) and an otherwise isogenic strain lacking the gene for the global regulatory gene, integration host factor (IHF) (experimental). $n = 4$ for both control and experimental situations. The horizontal axis represents the mean μ of the logarithm of the expression levels, and the vertical axis shows the corresponding standard deviations (std $= \sigma$). The left column corresponds to raw data; the right column to regularized standard deviations using Equation 5.13. Window size is $ws = 101$ and $K = 10$. Data are from [20].

Simulations

We have used the Bayesian approach and CyberT to analyze a number of published and unpublished data sets. We have found that the Bayesian approach very often compares favorably to a simple fold approach or a straight *t*-test and partially overcomes deficiencies related to low replication in a statistically consistent way [6]. In every high-density array experiment

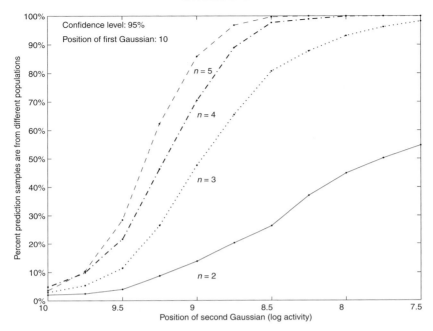

Figure 5.4. Plain *t*-test run on simulated data in the same range as data in previous figure. *n* points are randomly drawn from two pairs of Gaussians 1000 times and a plain *t*-test is applied to decide whether the two populations are different or not with a confidence of 95%. *n* is varied from 2 to 5. Vertical axis represents the percentage of "correct" answers. One of the Gaussians is kept fixed with a mean of -10 and standard deviation of 0.492. The other Gaussians are taken at regular intervals away from the first Gaussian. Both Gaussians have the corresponding average standard deviation derived by linear regression on the data of the previous figure in the interval $[-11, -7]$ of log-activities. The parameters of the regression line $\sigma = a \log$ (activity) $+ b$ for log-activities in the range of $[-11, -7]$ are: $a = -0.123$ and $b = -0.736$. For low values of *n*, the *t*-test performs very poorly.

we have analyzed, we have observed a strong scaling of the expression variance over replicated experiments with the average expression level, on both a log transformed and raw scale). As a result, a threshold for significance based solely on fold changes is likely to be too liberal for genes expressed at low levels and too conservative for highly expressed genes.

One informative data set for comparing the Bayesian approach to the simple *t*-test or fold change approaches is the high-density array experiment

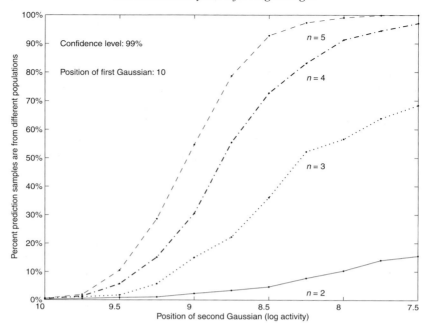

Figure 5.5. Same are previous figure but with 99% confidence interval.

reported in [20] comparing wild-type *Escherichia coli* cells to mutant cells for the global regulatory protein IHF (integration host factor). The main advantage of this data set is that it is fourfold replicated for both the wild-type and mutant alleles, according to the experimental design described in Figure 4.1. The regularizing effect of the prior based on the background standard deviation is shown in Figure 5.3 and in the simulation described below. The figure clearly shows that standard deviations vary substantially over the range of expression levels, in this case roughly in a monotonic decreasing fashion, although other behaviors also have been observed. Interestingly, in these plots the variance in log-transformed expression levels is higher for genes expressed at lower levels rather than at higher ones. These plots confirm that genes expressed at low or near background levels may require a stronger value of ν_0, or alternatively could be ignored in expression analyses. The variance in the measurement of genes expressed at a low level is large enough that in many cases it will be difficult to detect significant changes in expression for this class of loci.

In analyzing the data we found that large fold changes in expression are often associated with p-values not indicative of statistical change in the Bayesian framework, and conversely subtle fold changes are often highly significant as judged by the Bayesian analysis. In these two situations, the conclusions drawn with CyberT appear robust relative to those drawn from fold change alone, as large non-statistically significant fold changes are often associated with large measurement errors. As a result of the level of experimental replication seen in [20], we were able to look at the consistency of the Bayesian estimator of statistical significance relative to the t-test. We found that in independent samples of size two from the integration host factor (IHF) data set (i.e., two experiments versus two controls) the set of 120 most significant genes identified using the Bayesian approach had approximately 50% of their member in common, whereas the set of 120 most significant genes identified using the t-test had only approximately 25% of their members in common. This suggests that for twofold replication the Bayesian approach is approximately twice as consistent as a simple t-test at identifying genes as up- or down-regulated, although with only twofold replication there is a great deal of uncertainty associated with high-density array experiments (Figures 5.4 and 5.5).

To further assess the Bayesian approach, here we simulate an artificial data set assuming Gaussian distribution of log expressions, with means and variances in ranges similar to those encountered in the data set of [20], with 1000 replicates for each parameter combination. Selected means for the log data and associated standard deviations (in brackets) are as follows: -6 (0.1), -8 (0.2), -10 (0.4), -11 (0.7), -12 (1.0). On this artificially generated data, we can compare the behavior of a simple ratio (twofold and fivefold) approach, with a simple t-test, with the regularized t-test using the default settings of CyberT. The main results, reported in Table 5.1, can be summarized as follows:

- By 5 replications (5 control and 5 treatment) the Bayesian approach and t-test give similar results.
- When the number of replicates is "low" (2 or 3), the Bayesian approach performs better than the t-test.
- The false positive rate for the Bayesian and t-test approach are as expected (0.05 and 0.01 respectively) except for the Bayesian with very small replication (i.e., 2) where it appears elevated.
- The false positive rate on the ratios is a function of expression level and is much higher at lower expression levels. At low expression levels the false positive rate on the ratios is unacceptably high.

Table 5.1. *Number of positives detected out of 1000 genes*

n	Log expression from	to	Ratio 2-fold	5-fold	Plain t-test $p<0.05$	$p<0.01$	Bayes $p<0.05$	$p<0.01$
2	−8	−8	1	0	38	7	73	9
2	−10	−10	13	0	39	11	60	11
2	−12	−12	509	108	65	10	74	16
2	−6	−6.1	0	0	91	20	185	45
2	−8	−8.5	167	0	276	71	730	419
2	−10	−11	680	129	202	47	441	195
3	−8	−8	0	0	42	9	39	4
3	−10	−10	36	0	51	11	39	6
3	−12	−12	406	88	44	5	45	4
3	−6	−6.1	0	0	172	36	224	60
3	−8	−8.5	127	0	640	248	831	587
3	−10	−11	674	62	296	139	550	261
5	−8	−8	0	0	53	13	39	8
5	−10	−10	9	0	35	6	31	3
5	−12	−12	354	36	65	11	54	4
5	−6	−6.1	0	0	300	102	321	109
5	−8	−8.5	70	0	936	708	966	866
5	−10	−11	695	24	688	357	752	441
2v4	−8	−8	0	0	35	4	39	6
2v4	−10	−10	38	0	36	9	40	3
2v4	−12	−12	446	85	46	17	43	5
2v4	−6	−6.1	0	0	126	32	213	56
2v4	−8	−8.5	123	0	475	184	788	509
2v4	−10	−11	635	53	233	60	339	74

Notes:
Data was generated using normal distribution on a log scale in the range of [3],
with 1000 replicates for each parameter combination. Means of the log data and
associated standard deviations (in brackets) are as follows: −6 (0.1), −8 (0.2),
−10 (0.4), −11 (0.7), −12 (1.0). For each value of n, the first three experiments
correspond to the case of no change and therefore yield false positive rates.
Analysis was carried out using CyberT with default settings and the hyperparameter set to 10. Regularized t-tests were carried out using degrees of freedom equal
to: reps −1 + hyperparameter −1.

- For a given level of replication the Bayesian approach at $p<0.01$ detects
  more differences than a twofold change except for the case of low expression levels (where the false positive rate from ratios is elevated).
- The Bayesian approach with 2 replicates outperforms the t-test with 3
  replicates (or 2 versus 4 replicates).

- The Bayesian approach has a similar level of performance when comparing 3 treatments to 3 controls, or 2 treatments to 4 controls. This suggests an experimental strategy where the controls are highly replicated and a number of treatments less highly replicated.

Extensions

In summary, the probabilistic framework for array data analysis addresses a number of current approach shortcomings related to small sample bias and the fact that fold differences have different significance at different expression levels. The framework is a form of hierarchical Bayesian modeling with Gaussian gene-independent models. While there can be no perfect substitute for experimental replication (see also [21]), in simulations and controlled replicated experiments we have shown that the approach has a regularizing effect on the data, that it compares favorably to a conventional *t*-test, or simple fold approach, and that it can partially compensate for the absence of replication. New methods discussed in Chapter 7 are being developed to statistically estimate the rates of false positives and false negatives by modeling the distribution of *p*-values using a mixture of Dirichlet distributions and leveraging the fact that, under the null assumption, this distribution is uniform [22].

Depending on goals and implementation constraints, the present framework can be extended in a number of directions. For instance, regression functions could be computed offline to establish the relationship between standard deviation and expression level and used to produce background standard deviations. Another possibility is to use adaptive window sizes to compute the local background variance, where the size of the window could depend, for instance, on the derivative of the regression function. In an expression range in which the standard deviation is relatively flat (e.g., between −8 and −4 in Figure 5.3), the size of the window is less relevant than in a region where the standard deviation varies rapidly (e.g., between −12 and −10). A more complete Bayesian approach could also be implemented.

The general approach described in this chapter also can be extended to more complex designs and/or designs involving gradients of an experimental variable and/or time series designs. Examples would include a design in which cells are grown in the presence of different stressors (urea, ammonia, oxygen peroxide), or when the molarity of a single stressor is varied (0, 5, 10 mmol). Generalized linear and non-linear models can be used in this context. The most challenging problem, however, is to extend the probabilistic framework towards the second level of analysis, taking into account possible interactions and dependencies amongst genes. Multivariate normal models, and their mixtures, could provide the starting probabilistic models for this level of analysis (see also section "Gaussian processes" in Appendix B).

With a multivariate normal model, for instance, μ is a vector of means and Σ is a symmetric positive definite covariance matrix with determinant $|\Sigma|$. The likelihood has the form

$$C|\Sigma|^{-n/2}\exp\left[-\frac{1}{2}\sum_{i=1}^{n}(X_i-\mu)^T\Sigma^{-1}(X_i-\mu)\right] \qquad (5.17)$$

where T denotes matrix transpositions. The conjugate prior, generalizing the normal scaled inverse gamma distribution, is based on the inverse Wishart distribution (Appendix B) which generalizes the scaled inverse gamma distribution and provides a prior on Σ. In analogy with the one-dimensional case, the conjugate prior is parameterized by (μ_0, Λ_0/λ_0, ν_0, Λ_0). Σ has an inverse Wishart distribution with parameters ν_0 and Λ_0^{-1} (Appendix B). Conditioned on Σ, μ has a multivariate normal prior $\mathcal{N}(\mu; \mu_0, \Sigma/\lambda_0)$. The posterior then has the same form, a product of a multivariate normal with an inverse Wishart, parameterized by (μ_n, Λ_n/λ_n, ν_n, Λ_n). The parameters satisfy:

$$\mu_n = \frac{\lambda_0}{\lambda_0 + n}\mu_0 + \frac{n}{\lambda_0 + n}m$$

$$\lambda_n = \lambda_0 + n$$

$$\nu_n = \nu_0 + n$$

$$\Lambda_n = \Lambda_0 + \sum_1^n (X_i - m)(X_i - m)^t$$

$$+ \frac{\lambda_0 n}{\lambda_0 + n}(m - \mu_0)(m - \mu_0)^t \qquad (5.18)$$

Thus estimates similar to Equation 5.13 can be derived for this multidimensional case to regularize both variances and covariances. While multivariate normal and other related models may provide a good starting-point, good probabilistic models for higher-order effects in array data are still at an early stage of development. Many approaches so far have concentrated on more or less *ad hoc* applications of clustering methods. This is one of the main topics of Chapter 6 and 7. In this chapter we hope to have provided convincing argument to the reader that it is effective in general to model array data in probabilistic fashion. Besides DNA arrays, there are several other kinds of high-density arrays, at different stages of development, which could benefit from a similar treatment. Going directly to a systematic probabilistic framework may contribute to the accleration of the discovery process by avoiding some of the pitfalls observed in the history of sequence analysis, where it took several decades for probabilistic models to emerge as the proper framework for many tasks.

REFERENCES

1. Zhang, M. Q. Large-scale gene expression data analysis: a new challenge to computational biologists. 1999. *Genome Research* 9:681–688.
2. Beissbarth, T., Fellenberg, K., Brors, B., Arribas-Prat, R., Boer, J. M., Hauser, N. C., Scheideler, M, Hoheisel, J. D., Schutz, G., Poustka, A., and Vingron, M. Processing and quality control of DNA array hybridization data. 2000. *Bioinformatics* 16:1014–1022.
3. Zimmer, R., Zien, A., Aigner, T., and Lengauer, T. Centralization: a new method for the normalization of gene expression data. 2001. *Bioinformatics* 17 (Supplement 1): S323–S331.
4. Leach, S. M., Hunter, L., Taylor, R. C., and Simon, R. GEST: a gene expression search tool based on a novel bayesian similarity metric. 2001. *Bioinformatics* 17 (Supplement 1): S115–S122.
5. Baldi, P., and Long, A.D. A Bayesian framework for the analysis of microarray expression data: regularized t-test and statistical inferences of gene changes. 2001. *Bioinformatics* 17:509–519.
6. Long, A. D., Mangalam, H. J., Chan, B. Y., Tolleri, L., Hatfield, G. W., and Baldi, P. Global gene expression profiling in *Escherichia coli* K12: improved statistical inference from DNA microarray data using analysis of variance and a Bayesian statistical framework. 2001. *Journal of Biological Chemistry* 276:19937–19944.
7. Schena, A. M., Shalon, D., Davis, R. W., and Brown, P. O. Quantitative monitoring of gene expression patterns with a complementary DNA microarray. 1995. *Science* 270:467–470.
8. Schena, B. M., Shalon, D., Heller, R., Chai, A., Brown, P. O. and Davis, R. W. Parallel human genome analysis: microarray-based expression monitoring of 1000 genes. 1995. *Proceedings of the National Academy of Sciences of the USA* 93:10614–10619.
9. Heller, C. R. A., Schena, M., Chai, A., Shalon, D., Bedillon, T., Gilmore, J., Wolley, D. E., and Davis, R. W. Discovery and analysis of inflammatory disease-related genes using cDNA microarrays. 1997. *Proceedings of the National Academy of Sciences of the USA* 94:2150–2155.
10. Baldi, P., and Brunak, S. *Bioinformatics: The Machine Learning Approach*, 2nd edn. 2001. MIT Press, Cambridge, MA.
11. Durbin, R., Eddy, S., Krogh, A., and Mitchison, G. *Biological Sequence Analysis: Probabilistic Models of Proteins and Nucleic Acids*. 1998. Cambridge University Press, Cambridge.
12. Box, G. E. P., and Tiao, G. C. *Bayesian Inference in Statistical Analysis*. 1973. Addison Wesley, New York.

13. Berger, J. O. *Statistical Decision Theory and Bayesian Analysis.* 1985. Springer-Verlag, New York.
14. Pratt, J. W., Raiffa, H., and Schlaifer, R. *Introduction to Statistical Decision Theory.* 1995. MIT Press, Cambridge, MA.
15. Gelman, A., Carlin, J. B., Stern, H. S., and Rubin, D. B. *Bayesian Data Analysis.* 1995. Chapman & Hall, London.
16. Wiens, B. L. When log-normal and gamma models give different results: a case study. 1999. *American Statistician* 53:89–93.
17. Newton, M. A., Kendziorski, C. M., Richmond, C. S., Blattner, F. R., and Tsui, K. W. On differential variability of expression ratios: improving statistical inference about gene expression changes from microarray data. In press. *Journal of Computational Biology.*
18. MacKay, D. J. C. Bayesian interpolation. 1992. *Neural Computation* 4:415–447.
19. MacKay, D. J. C. Comparison of approximate methods for handling hyperparameters. 1999. *Neural Computation* 11:1035–1068.
20. Arfin, S. M., Long, A.D., Ito, E.T., Tolleri, L., Riehle, M. M., Paegle, E. S., and Hatfield, G. W. Global gene expression profiling in *Escherichia coli* K12: the effects of integration host factor. 2000. *Journal of Biological Chemistry* 275:29672–29684.
21. Lee, M. T., Kuo, F. C., Whitmore, G. A., and Sklar, J. Importance of replication in microarray gene expression studies: statistical methods and evidence from repetitive cDNA hydribizations. 2000. *Proceedings of the National Academy of Sciences of the USA* 97:9834–9839.
22. Allison, D. B., Gadbury, G. L., Moonseong, H., Fernandez, J. R., Cheol-Koo, L., Prolle, T. A., and Weindruch, R. A mixture model approach for the analysis of microarray gene expression data. 2002. *Computational Statistics and Data Analysis* 39:1–20.

6

Statistical analysis of array data: Dimensionality reduction, clustering, and regulatory regions

Problems and approaches

Differential expression is a useful tool for the analysis of DNA microarray data. However, and in spite of the fact that it can be applied to a large number of genes, differential analysis remains within the confines of the old one-gene-at-a-time paradigm. Knowing that a gene's behavior has changed between two situations is at best a first step. In a cancer experiment, for instance, a significant change could be associated with a direct causal link (activation of an oncogene), a more indirect chain of effects (signaling pathway), a non-specific related phenomenon (cell division), or even a spurious event completely unrelated to cancer ("noise").

Most, if not all, genes act in concert with other genes. What DNA microarrays are really after are the *patterns* of expression across multiple genes and experiments. And to detect such patterns, additional methods such as clustering must be introduced. In fact, in the limit, differential analysis can be viewed as a clustering method with only two clusters: change and no-change. Thus, at the next level of data analysis, we want to remove the simplifying assumption that genes are independent and look at their covariance, at whether there exist multi-gene patterns, clusters of genes that share the same behavior, and so forth. While array data sets and formats remain heterogeneous, a key challenge in time is going to be the development of methods that can extract order across experiments, in typical data sets of size 30000×1000 and model, for instance, the statistical distribution of a gene's expression levels over the space of possible conditions. Not surprisingly, conceptually these problems are not completely unlike those encountered in population genetics, such as detecting combinations of SNPs associated with complex diseases.

The methods and examples of Chapter 5 and 7 show how noisy the boundaries between clusters can be, especially in isolated experiments with low repetition. The key observation, however, is that even when individual experiments are not replicated many times complex expression patterns can still be detected robustly across multiple experiments and conditions. Consider, for instance, a cluster of genes directly involved in the cell division cycle and whose expression pattern oscillates during the cycle. For each individual measurement at a given time t, noise alone can introduce distortions so that a gene which belongs to the cluster may fall out of the cluster. However, when the measurements at other times are also considered, the cluster becomes robust and it becomes unlikely for a gene to fall out of the cluster it belongs to at most time steps. The same can be said of course for genes involved in a particular form of cancer across multiple patients, and so forth. In fact, it may be argued (Chapter 8) that robustness is a fundamental characteristic of regulatory circuits that must somehow transpire even through noisy microarray data.

In many cases, cells tend to produce the proteins they need simultaneously, and only when they need them. The genes for the enzymes that catalyze a set of reactions along a pathway are likely to be co-regulated (and often somewhat co-located along the chromosome). Thus, depending on the data and clustering methods, gene clusters can often be associated with particular pathways and with co-regulation. Even partial understanding of the available information can provide valuable clues. Co-expression of novel genes may provide a simple means of gaining leads to the functions of many genes for which information is not yet available. Likewise, multi-gene expression patterns could characterize diseases and lead to new precise diagnostic tools capable of discriminating, for instance, different kinds of cancers.

Many data analysis techniques have already been applied to problems in this class, including various clustering methods from k-means to hierarchical clustering, principal component analysis, factor analysis, independent component analysis, self-organizing maps, decision trees, neural networks, support vector machines, and Bayesian networks to name just a few. It is impossible to review all the methods of analysis in detail in the available space and counter-productive to try to single out a "best method" because: (1) each method may have different advantages depending on the specific task and specific properties of the data set being analyzed; (2) the underlying technology is still rapidly evolving; and (3) noise levels do not always allow for a fine discrimination between methods. Rather, we focus on the main methods of analysis and the underlying mathematical background.

Array data is inherently high-dimensional, hence methods that try to reduce the dimensionality of the data and/or lend themselves to some form of visualization remain particularly useful. These range from simple plots of one condition versus another, to projection on to lower dimensional spaces, to hierarchical and other forms of clustering. In the next sections, we focus on dimensionality reduction (principal component analysis) and clustering since these are two of the most important and widely used methods of analysis for array data. For clustering, we partially follow the treatment in [1]. Additional information about other methods of array data analysis can be found in the last chapter (neural networks and Bayesian networks), in the Appendices (support vector machines), and throughout the references. Clustering methods of course can be applied not only to genes, but also to conditions, DNA sequences, and other relevant data. From array-derived gene clusters it is also possible to go back to the corresponding gene sequences and look, for instance, for shared motifs in the regulatory regions of co-regulated genes, as described in the last section of this chapter.

Visualization, dimensionality reduction, and principal component analysis

The simplest approach to visually explore current array data is perhaps a two-dimensional plot of the activity levels of the genes in one experimental condition versus another. When each experiment is repeated several times, the average values can be used. In such plots, typically most genes are found along the main diagonal (assuming similar calibration between experiments) while differentially expressed genes appear as outliers. The methods described in Chapter 5 allow quantitative analysis of these outliers based on individual experimental values.

A second more sophisticated approach for dimensionality reduction and visualization is principal component analysis. Principal component analysis (PCA) is a widely used statistical data analysis technique that can be viewed as: (1) a method to discover or reduce the dimensionality of the data; (2) a method to identify new meaningful underlying variables; (3) a method to compress the data; and (4) a method to visualize the data. It is also called the Karhunen–Loéve transform. Hotelling transform, or singular value decomposition (SVD) and provides an optimal linear dimension reduction technique in the mean-square sense.

Consider a set of N points $x_1,..., x_N$ in a space of dimension M. In the case of array data, the points could correspond to genes if the axes are associated with different experiments or to experiments if the axes are associated with

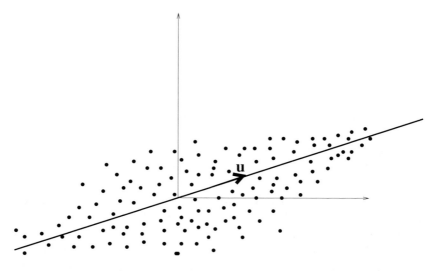

Figure 6.1. A two-dimensional set of points with its principal component axes and its unit vector u.

the probes/genes. We assume that the xs have already been centered by subtracting their mean value, or expectation, $E[x]$. The basic idea in PCA is to reduce the dimension of the data by projecting the xs on to an interesting linear subspace of dimension K, where K is typically significantly smaller than M. Interesting is defined in terms of variance maximization.

PCA can easily be understood by recursively computing the orthonormal axis of the projection space. For the first axis, we are looking for a unit vector u_1 such that, on average, the squared length of the projection of the xs along u_1 is maximal. Assuming that all vectors are column vectors, this can be written as

$$u_1 = \arg \max_{\|u\|=1} E[(u^T x)^2] \tag{6.1}$$

where u^T denotes transposition, and E is the expected or average value (Figure 6.1). Any vector is always the sum of its orthogonal projections on to a given subspace and the orthogonal complement, so that here $x = (u^T x)u + (x - (u^T x)u)$. The second component maximizes the residual variance associated with $(x = (u^T x)u)$ and so forth. More generally, if the first $u_1, ..., u_{k-1}$ components have been determined, the next component is the one that maximizes the residual variance in the form

$$u_k = \arg \max_{\|u\|=1} E\left[\left(x - \sum_{i=1}^{k-1} (u_i^T x)u_i \right)^2 \right] \tag{6.2}$$

The principal components of the vector x are given by $c_i = u_i^T x$. By construction, the vectors u_i are orthonormal. In practice, it can be shown the u_is are the eigenvectors of the (sample) covariance matrix $\Sigma = E[xx^T]$ associated with the K largest eigenvalues and satisfy

$$\Sigma u_k = \lambda_k u_k \qquad (6.3)$$

In array experiments, these give rise to "eigengenes" and "eigenarrays" [2]. Each eigenvalue λ_k provides a measure of the proportion of the variance explained by the corresponding eigenvector.

By projecting the vectors onto the subspace spanned by the first eigenvectors, PCA retains the maximal variance in the projection space and minimizes the mean-square reconstruction error. The choice of the number K of components is in general not a serious problem – basically it is a matter of inspecting how much variance is explained by increasing values of K. For visualization purposes, only projections on to two- or three-dimensional spaces are useful. The first dominant eigenvectors can be associated with the discovery of important features or patterns in the data. In DNA microarray data where the points correspond to genes and the axes to different experiments, such as different points in time, the dominant eigenvectors can represent expression patterns. For example, if the first eigenvector has a large component along the first experimental axis and a small component along the second and third axis, it can be associated with the experimental expression pattern "high–low–low". In the case of replicated experiments, we can expect the first eigenvector to be associated with the principal diagonal $\left(1/\sqrt{N}, \ldots, 1/\sqrt{N}\right)$.

There are also a number of techniques for performing approximate PCA, as well as probabilistic and generalized (non-linear and project pursuit) versions of PCA [3, 4, 5, 6, 7], and references therein. An extensive application of PCA techniques to array data is described in [2] and in Chapter 7.

Although PCA is not a clustering technique *per se*, projection on to lower dimensional spaces associated with the top components can help reveal and visualize the presence of clusters in the data (see Chapter 7 for an example). These projections however must be considered carefully since clusters present in the data can become hidden during the projection operation (Figure 6.2). Thus, while PCA is a useful technique, it is only one way of analyzing the data which should be complemented by other methods, and in particular by methods whose primary focus is data clustering.

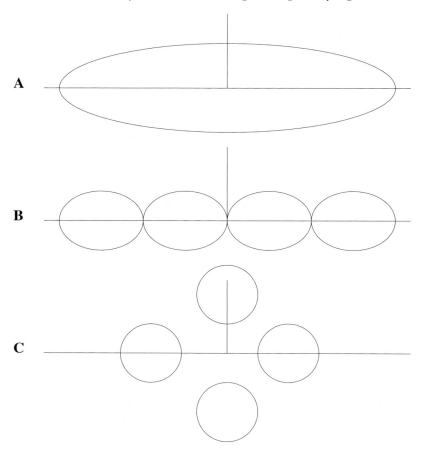

Figure 6.2. Schematic representation of three data sets in two dimensions with very different clustering properties and projections onto principal components. A has one cluster. B and C have four clusters. A, B, and C have the same principal components. A and B have a similar covariance matrix with a large first eigenvalue. The first and second eigenvalues of C are identical.

Clustering overview

Another direction for visualizing or compressing high-dimensional array data is the application of clustering methods. Clustering refers to an important family of techniques in exploratory data analysis and pattern discovery, aimed at extracting underlying cluster structures. Clustering, however, is a "fuzzy" notion without a single precise definition. Dozens of clustering algorithms exist in the literature and a number of *ad hoc* clustering procedures, ranging from hierarchical clustering to k-means have been applied to

DNA array data [8, 9, 10, 11, 12, 13, 14, 15]. Because of the variety and "open" nature of clustering problems, it is unlikely that a systematic exhaustive treatment of clustering can be given. However there are a number of important general issues to consider in clustering and clustering algorithms, especially in the context of gene expression.

Data types

At the highest level, clustering algorithms can be distinguished depending on the nature of the data being clustered. The standard case is when the data points are vectors in Euclidean space. But this is by no means the only possibility. In addition to vectorial data, or numerical data expressed in absolute coordinates, there is the case of relational data, where data is represented in relative coordinates, by giving the pairwise distance between any two points. In many cases the data is expressed in terms of a pairwise similarity (or dissimilarity) measure that often does not satisfy the three axioms of a distance (positivity, symmetry, and triangle inequality). There exist situations where data configurations are expressed in terms of ternary or higher order relationships or where only a subset of all the possible pairwise similarities is given. More importantly, there are cases where the data is not vectorial or relational in nature, but essentially qualitative, as in the case of answers to a multiple-choice questionnaire. This is sometimes also called nominal data. While at the present time gene expression array data is predominantly numerical, this is bound to change in the future. Indeed, the dimension "orthogonal to the genes" covering different experiments, different patients, different tissues, different times, and so forth is at least in part non-numerical. As databases of array data grow, in many cases the data will be mixed with both vectorial and nominal components.

Supervised/unsupervised

One important distinction amongst clustering algorithms is supervised versus unsupervised. In supervised clustering, clustering is based on a set of given reference vectors or classes. In unsupervised clustering, no predefined set of vectors or classes is used. Hybrid methods are also possible where an unsupervised approach is followed by a supervised one. At the current early stage of gene expression array experiments, unsupervised methods such as k-means and self-organizing maps [11] are most commonly used. However, supervised methods have also been tried [12, 16], where clusters are predetermined using functional information or unsupervised clustering methods, and then new genes are classified in the various clusters using a classifier, such as linear and quadratic discriminant analysis, decision trees, neural networks, or support vector machines, that can learn the decision boundaries between data classes. The feasibility of class discrimination with array expression data has been demonstrated, for instance for tumor classes such as leukemias arising from several different precursors [17], and B-cell lymphomas [18] (see also [19, 20]).

Similarity/distance

The starting-point of several clustering algorithms, including several forms of hierarchical clustering, is a matrix of pairwise similarities or distances between the objects to be clustered. In some instances, this pairwise distance is replaced by a distortion measure between a data point and a class centroid as in vector quantization methods. The precise definition of similarity, distance, or distortion is crucial and, of course, can greatly impact the output of the clustering algorithm. In any case, it allows converting the clustering problem into an optimization problem in various ways, where the goal is essentially to find a relatively small number of classes with high intraclass similarity or low intraclass distortion, and good interclass separation. In sequence analysis, for instance, similarity can be defined using a score matrix for gaps and substitutions and an alignment algorithm. In gene expression analysis, different measures of similarity can be used. Two obvious examples are Euclidean distance (or more generally L^p distances) and correlation between the vectors of expression levels. The Pearson correlation coefficient is just the dot product of two normalized vectors, or the cosine of their angle. It can be measured on each pair of genes across, for instance, different experiments or different time steps. Each measure of similarity comes with its own advantages and drawbacks depending on the situation, and may be more or less suitable to a given analysis. The correlation, for instance, captures similarity in shape but places no emphasis on the magnitude of the two series of measurements and is quite sensitive to outliers. Consider, for instance, measuring the activity of two unrelated genes that are fluctuating close to the background level. Such genes are very similar in Euclidean distance (distance close to 0), but dissimilar in terms of correlation (correlation close to 0). Likewise, consider the two vectors 1000000000 and 0000000001. In a sense they are similar since they are almost always identical and equal to 0. On the other hand, their correlation is close to 0 because of the two "outliers" in the first and last position.

Number of clusters

The choice of the number K of clusters is a delicate issue, which depends, among other things, on the scale at which one looks at the data. It is safe to say that an educated partly manual trial-and-error approach still remains an efficient and widely used technique, and this is true for array data at the present stage. Because in general the number of clusters is relatively small, all possible values of K within a reasonable range can often be tried. Intuitively, however, it is clear that one ought to be able to assess the quality of K from the compactness of each cluster and how well each cluster is separated from the others. Indeed there have been several recent developments aimed at the automatic determination of the number of clusters [13, 21, 22, 23] with reports of good results.

Cost function and probabilistic interpretation

Any rigorous discussion of clustering on a given data set presupposes a principled way of comparing different ways of clustering the same data, hence the need for some kind of global cost/error function that can easily be computed. The goal of clustering then is to try to minimize such function. This is also called parametric clustering in the literature, as opposed to non-parametric clustering, where only local functions are available [24].

In general, at least for numerical data, this function will depend on quantities such as the centers of the clusters, the distance of each point in a cluster to the corresponding center, the average degree of similarity of the points in a given cluster, and so forth. Such a function is often discontinuous with respect to the underlying clustering of the data. Here again there are no universally accepted functions and the cost function should be tailored to the problem, since different cost functions can lead to different answers.

Because of the advantages associated with probabilistic methods and modeling, it is tempting to associate the clustering cost function with the negative log-likelihood of an underlying probabilistic model. While this is formally always possible, it is of most interest when the structure of the underlying probabilistic model and the associated independence assumptions are clear. This is when the additive terms of the cost function reflect the factorial structure of the underlying probabilities and variables. As we shall see this is the case with mixture models, where the k-means clustering algorithm can be viewed as a form of expectation maximization (EM).

In the rest of this section, we describe in more detail two basic clustering algorithms that can be applied to DNA array data, hierarchical clustering and k-means. Many other related approaches, including vector quantization [25], graph methods [13], factorial analysis, and self-organizing maps can be found in the references.

Hierarchical clustering

Clusters can result from a hierarchical branching process. Thus there exist methods for automatically building a tree from data given in the form of pairwise similarities. In the case of gene expression data, this is the approach used in [8].

Hierarchical clustering algorithm

The standard algorithm used in [8] recursively computes a dendrogram that assembles all the elements into a tree, starting from the correlation (or distance or similarity) matrix C. The algorithm starts by assigning a leaf of the tree to each element (gene). At each step of the algorithm:

- The two most similar elements of the current matrix (highest correlation) are computed and a node joining these two elements is created.
- An expression profile (or vector) is created for the node by averaging the two expression profiles (or vectors) associated with the two points (missing data can be ignored and the average can be weighted by the number of elements they contain). Alternatively, a weighted average of the distances is used to estimate the new distance between centers without actually computing the profiles.
- A new smaller correlation matrix is computed using the newly computed expression profile or vector and replacing the two joined elements with the new node.
- With N starting points, the process is repeated at most $N-1$ times, until a single node remains.

This algorithm is familiar to biologists and has been used in sequence analysis, phylogenetic trees, and average-linkage cluster analysis. As described, it requires $O(N^3)$ steps since for each of the $N-1$ fusions one must search for an optimal pair. An $O(N^2)$ version of the algorithm is described in [26]. The output of hierarchical clustering is typically a binary tree and not a set of clusters. In particular, it is usually not obvious how to define clusters from the tree since clusters are derived by cutting the branches of the tree at more or less arbitrary points.

Tree visualization

In the case of gene expression data, the resulting tree organizes genes or experiments so that underlying biological structure can often be detected and visualized [8, 9, 18, 27]. As already pointed out, after the construction of such a dendrogram there is still a problem of how to display the result and which clusters to choose. Leaves are often displayed in linear order and biological interpretations are often made in relation to this order, e.g., adjacent genes are assumed to be related in some fashion. Thus the order of the leaves matters.

At each node of the tree, either of the two elements joined by the node can be ordered to the left or the right of the other. Since there are $N-1$ joining steps, the number of linear orderings consistent with the structure of the tree is 2^{N-1}. Computing the optimal linear ordering maximising the combined similarity of all neighboring pairs seems difficult, and therefore heuristic approximations have been proposed [8]. These approximations weigh genes using average expression level, chromosome position, and time of maximal induction.

More recently, it was noticed in [28] that the optimal linear ordering can be computed in $O(N^4)$ steps simply by using dynamic programming, in a form which is essentially the well-known inside portion of the inside–outside algorithm for stochastic context-free grammars [1]. If $G_1,\ldots,$ G_N are the leaves of the tree and ϕ denotes one of the 2^{N-1} possible orderings of the leaves, we would like to maximize

$$\sum_{i=1}^{N-1} C[G_{\phi(i)}, G_{\phi(i+1)}] \qquad (6.4)$$

where $G_{\phi(i)}$ is the ith leaf when the tree is ordered according to ϕ. Let V denote both an internal node of the tree as well as the corresponding subtree. V has two children: V_l on the left and V_r on the right, and four grandchildren V_{ll}, V_{lr}, V_{rl}, and V_{rr}. The algorithm works bottom up, from the leaves towards the roots by recursively computing the cost of the optimal ordering $M(V,U,W)$ associated with the subtree V when U is the leftmost leaf of V_l and W is the rightmost leaf of V_r (Figure 6.3). The dynamic programming recurrence is given by:

$$M(V,U,W) = \max_{R\in V_{lr}, S\in V_{rl}} M(V_l,U,R) + M(V_r,S,W) + C(R,S) \qquad (6.5)$$

The optimal cost $M(V)$ for V is obtained by maximizing over all pairs, U, W. The global optimal cost is obtained recursively when V is the root of the tree, and the optimal tree can be found by standard backtracking. The algorithm requires computing $M(V,U,W)$ only once for each $O(N^2)$ pair of leaves. Each computation of $M(V,U,W)$ requires maximization over all possible $O(N^2)$ (R,S) pair of leaves. Hence the algorithm requires $O(N^4)$ steps with $O(N^2)$ space complexity, since only one $M(V,U,W)$ must be computed for each pair (U,W) and this is also the size of the pairwise similarity

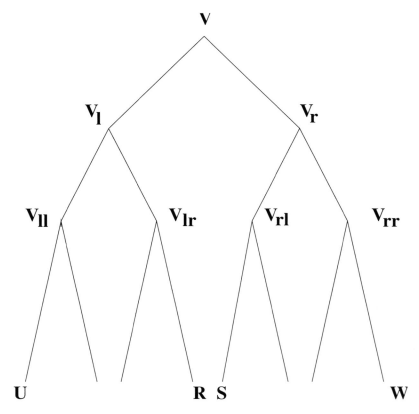

Figure 6.3. Tree underlying the dynamic programming recurrence of the inside algorithm.

matrix C. For some applications, $O(N^4)$ is too slow. A faster algorithm on average is developed in [28] by early termination of search paths that are not promising.

   Hierarchical clustering has proved to be a useful for array data analysis in the literature, for instance for finding genes that share a common function [8, 18, 27]. The main clusters derived are often biologically significant and the optimal leaf ordering algorithm can further improve the quality and interpretability of the results [28]. Optimal leaf ordering helps in improving the definition of cluster boundaries and the relationships between clusters.

K-means, mixture models, and EM algorithms

K-means algorithm

Of all clustering algorithms, k-means [29] is among the simplest and most widely used, and has probably the cleanest probabilistic interpretation as a form of EM on the underlying mixture model. In a typical implementation of the k-means algorithm, the number of clusters is fixed to some value of *K* based, for instance, on the expected number of regulatory patterns. *K* representative points or centers are initially chosen for each cluster more or less at random. In array data, these could reflect, for instance, the expected number of regulatory patterns. These points are also called centroids or prototypes. Then at each step:

- Each point in the data is assigned to the cluster associated with the closest representative.
- After the assignment, new representative points are computed for instance by averaging or taking the center of gravity of each computed cluster.
- The two procedures above are repeated until the system converges or fluctuations remain small.

Hence notice that using k-means requires choosing the number of clusters and also being able to compute a distance or similarity between points and compute a representative for each cluster given its members.

The general idea behind k-means can lead to different software implementations depending on how the initial centroids are chosen, how symmetries are broken, whether points are assigned to clusters in a hard or soft way, and so forth. A good implementation ought to run the algorithm multiple times with different initial conditions and possibly also try different values of *K* automatically.

When the cost function corresponds to an underlying probabilistic mixture model [30, 31], k-means is an online approximation to the classical EM algorithm [1, 32], and as such in general is bound to converge towards a solution that is at least a local maximum likelihood or maximum posterior solution. A classical case is when Euclidean distances are used in conjunction with a mixture of Gaussians model. A related application to a sequence clustering algorithm is described in [33].

Mixture models and EM

To better understand the connection to mixture models, imagine a data set $D = (d_1,\ldots,d_N)$ and an underlying mixture model with *K* components of the form

$$P(d) = \sum_{k=1}^{K} P(M_k)P(d\,|\,M_k) = \sum_{k=1}^{K} \lambda_k P(d\,|\,M_k) \tag{6.6}$$

where $\lambda_k \geq 0$ and $\sum_k \lambda_k = 1$ and M_k is the model for cluster k. Mixture distributions provide a flexible way for modeling complex distributions, combining together simple building-blocks, such as Gaussian distributions. The Lagrangian associated with the log-likelihood and the normalization constraints on the mixing coefficients is given by

$$\mathcal{L} = \sum_{i=1}^{N} \log\left(\sum_{k=1}^{K} \lambda_k P(d_i\,|\,M_k) \right) - \mu\left(\sum_{k=1}^{K} \lambda_k - 1 \right) \tag{6.7}$$

with the corresponding critical equation

$$\frac{\partial \mathcal{L}}{\partial \lambda_k} = \sum_{i=1}^{N} \frac{P(d_i\,|\,M_k)}{P(d_i)} - \mu = 0 \tag{6.8}$$

Multiplying each critical equation by λ_k and summing over k immediately yields the value of the Lagrange multiplier $\mu = N$. Multiplying again the critical equation across by $P(M_k) = \lambda_k$, and using Bayes' theorem in the form

$$P(M_k\,|\,d_i) = P(d_i\,|\,M_k)P(M_k)/P(d_i) \tag{6.9}$$

yields

$$\lambda_k^* = \frac{1}{N} \sum_{i=1}^{N} P(M_k\,|\,d_i) \tag{6.10}$$

Thus the maximum likelihood estimate of the mixing coefficients for class k is the sample mean of the conditional probabilities that d_i comes from model k. Consider now that each model M_k has its own vector of parameters (w_{kj}). Differentiating the Lagrangian with respect to w_{kj} gives

$$\frac{\partial \mathcal{L}}{\partial w_{kj}} = \sum_{i=1}^{N} \frac{\lambda_k}{P(d_i)} \frac{\partial P(d_i\,|\,M_k)}{\partial w_{kj}} \tag{6.11}$$

Substituting Equation 6.9 in Equation 6.11 finally provides the critical equation

$$\sum_{i=1}^{N} P(M_k\,|\,d_i) \frac{\partial \log P(d_i\,|\,M_k)}{\partial w_{kj}} = 0 \tag{6.12}$$

for each k and j. The maximum likelihood equations for estimating the parameters are weighted averages of the maximum likelihood equations

$$\partial \log P(d_i\,|\,M_k)/\partial w_{kj} = 0 \tag{6.13}$$

arising from each point separately. As in Equation 6.10, the weights are the probabilities of membership of the d_i in each class.

The maximum likelihood Equations 6.10 and 6.12 can be used iteratively to search for maximum likelihood estimates; this is in essence the EM algorithm. In the E step, the membership probabilities (hidden variables) of each data point are estimated for each mixture component. The M step is equivalent to K separate estimation problems with each data point contributing to the log-likelihood associated with each of the K components with a weight given by the estimated membership probabilities. Different flavors of the same algorithm are possible depending on whether the membership probabilities $P(M|d)$ are estimated in hard or soft fashion during the E step. The description of k-means given above corresponds to the hard version where these membership probabilities are either 0 or 1, each point being assigned to only one cluster. This is analogous to the use of the Viterbi version of the EM algorithm for hidden Markov models, where only the optimal path associated with a sequence is used, rather than the family of all possible paths. Different variations are also possible during the M step of the algorithms depending, for instance, on whether the parameters w_{kj} are estimated by gradient descent or by solving Equation 6.12 exactly. It is well known that the center of gravity of a set of points minimizes its average quadratic distance to any fixed point. Therefore in the case of a mixture of spherical Gaussians, the M step of the k-means algorithm described above maximizes the corresponding quadratic log-likelihood and provides a maximum likelihood estimate for the center of each Gaussian component. It is also possible to introduce prior distributions on the parameters of each cluster and/or the mixture coefficients and create more complex hierarchical mixture models.

PCA, hierarchical clustering, k-means, as well as other clustering and data analysis algorithms are currently implemented in several publicly or commercially (Table 6.1) available software packages for DNA array data analysis. It is important to recognize that many software packages will output some kind of answer, for instance a set of clusters, on any kind of data set. These answers should not be trusted always blindly. Rather it is wise practice, whenever possible, to track down the assumptions underlying each algorithm/implementation/package and to run the same algorithms with different parameters, as well as different algorithms, on the same data set, as well as on other data sets.

Table 6.1. *Commercially available software packages for DNA array data analysis*

Program	Platform	Provider name	Web address	Types of functions
Commercial sources				
GeneSpring 4.1	Win Mac Unix	Silicon Genetics	www.sigenetics.com	*t*-test, hierarchical clustering, k-means, SOM, PCA, class predictor; experiment tree
Spotfire	Win Unix	Spotfire	www.spotfire.com	*t*-test, PCA, k-means, hierarchical clustering
GeneSight 2.1	Win	Biodiscovery	www.biodiscovery.com	*t*-test, k-means, hierarchical clustering, SOM, PCA, pattern similarity search, non-linear normalization
Data Mining Tool	Win	Affymetrix	www.affymetrix.com	*t*-test, Mann–Whitney test, SOM, modified Pearson's correlation coefficient
Web-based sources				
GeneMaths	Win	Applied Maths	www.applied-maths.com	Hierarchical clustering, k-means, SOM, PCA, discrimination analysis with or without variance
CyberT	Web-based	UCI	www.genomics.uci.edu	*t*-test or *t*-test with Bayesian framework
R Cluster	Web-based	UCI	www.genomics.uci.edu	Hierarchical clustering, k-means

Name	Platform	Source	URL	Methods
EPCLUST/ means Expression Profiler	Web-based	European Bioinformatic Institute	www.ebi.ac.uk/microarray	Hierarchical clustering, k-means and finding the nearest neighbors
Freeware sources				
J-Express	Mac Win Unix	Molmine	www.molmine.com	Hierarchical clustering, k-means, PCA, SOM, profile similarity search
Significance analysis of microarray (SAM)	Win	Stanford University	www-stat-class.stanford.edu/SAM/ SAMServlet	
dChip	Win	Harvard University (Wong)	www.biostat.harvard.edu/ complab/dchip/dchip.exe	t-test, model-based analysis for GeneChips™
Cluster and TreeView	Win	Stanford University/ Berkeley (Eisen)	rana.lbl.gov/Eisen Software.htm	Hierarchical clustering, k-means, PCA, SOM
SVM 1.2	Linux Unix	Columbia University (Noble)	www.cs.columbia.edu/ ~noble/svm/doc/	Class prediction
Xcluster	Mac Win Unix	Stanford University (Sherlock)	genome-www.stanford. edu/~sherlock/cluster.Html	SOM, k-means
GeneCluster	Win	Whitehead Institute/MIT	http://www.genome.wi.mit. edu/MPR	SOM

DNA arrays and regulatory regions

Another important level of analysis consists in combining DNA array data with DNA sequence data, and in particular with regulatory regions. This combination can be used to detect regulatory motifs, but also to address global questions of regulation. While gene regulatory elements have been found in a variety of regions including introns, distant intragenic regions, and downstream regions, the bulk of the regulation of a given gene is, in general, believed to depend primarily on a more or less extended region immediately upstream of the gene. In [34], for instance, this model was tested on a genomic scale by coupling expression data obtained during oxidative stress response with all pairwise alignments of yeast ORF upstream regions. In particular, it was found that as the difference in upstream regions increases, the correlation in activity rapidly drops to zero and that divergent ORFs, with overlapping upstream regions, do not seem to have correlated expression levels. By and large, however, the majority of current efforts aimed at combining DNA array and sequence data are focused on searching for regulatory motifs.

Several techniques have been developed for the discovery of "significant" patterns from a set of unaligned DNA sequences. Typically these patterns represent regulatory (transcription factor DNA binding sites) or structural motifs that are shared in some form by the sequences. The length and the degeneracy of the pattern are of course two important parameters [35, 36, 37]. Probabilistic algorithms such as EM and Gibbs sampling naturally play an essential role in motif finding, due to both the structural and location variability of motifs (see programs such as MEME and CONSENSUS).

Simple measures of overrepresentation also have been shown to be effective for detecting such motifs, for instance in sets of gene-upstream or gene-downstream [38] regions. While these data mining algorithms can be applied using a purely combinatorial approach to genomic DNA [39, 40], the methods and results can be further refined, and the sensitivity increased, by focusing the search on specific clusters of genes derived from array data analysis, such as clusters of genes that appear to be co-regulated. In addition to regulatory motifs found in the TRANSFAC database [41], these methods can detect novel motifs in the large amounts of more or less unannotated genomic DNA that has become available through genome and other sequencing projects [34, 42, 43, 44, 45, 46].

The basic idea behind these approaches is to compute the number of occurrences of each k-mer, typically for values of k in the range of 3 to 10, within a set of sequences, such as all gene-upstream regions, or all the upstream regions of a particular set of co-regulated genes, and look for k-mers that are overrepresented. Overrepresentation is a statistical concept that can be assessed in a number of different ways and, depending on the problem, a number of points must be carefully considered. These include:

- Regions: How are the upstream, or downstream, regions defined? Do they have fixed length? How are they treated with respect to neighboring genes on each strand and possible overlaps?
- Counts: Are the two strands treated separately or aggregated? It is well known, for instance, that certain regulatory motifs are active regardless of the strand on which they occur and these are better detected if counts on both strands are aggregated. Other motifs are strand-specific.
- Background model: overrepresentation must be assessed with respect to a statistical background model. The choice of the background model is critical and non-trivial. In particular the background model cannot be too good otherwise it would predict the counts exactly and therefore would be worthless. Typical models used in the literature are Markov models of various orders, measured on the data or some other reference set. Another possible background model, is to consider the average of single (or multiple) base pair mismatches [45], i.e., to estimate the counts of a given k-mer using the counts of all the k-mers that differ in one position.
- Statistics: Several statistics can be used to detect significant overrepresentation from the raw counts, such as ratio, log-likelihood, z-score binomial, t-test, Poisson, and compound Poisson. As in the case of array data, inference based on ratio alone can be tricky, especially for low expected frequencies that can induce false positives (e.g., 1 versus 4 is very different from 1000 versus 4000).
- Gene clusters: If the method is applied to the DNA sequences associated with a cluster of genes derived from array data, how is the cluster determined? Are the genes up- or down-regulated under a given condition? Etc. Notice also that array data can be used as a filter to detect overrepresentation before or after the counts, often yielding somewhat different results.

Overrepresented k-mers are of particular interest and have been shown to comprise well-known regulatory motifs. For instance, when the algorithms are run on the yeast upstream regions using oxidative stress data, one immediately detects the well-known stress element CCCCT [47] and its reverse complement AGGGG, or the YAP1 element TTACTAA and its reverse complement TTAGTAA [48, 49, 50]. In general, however, only a

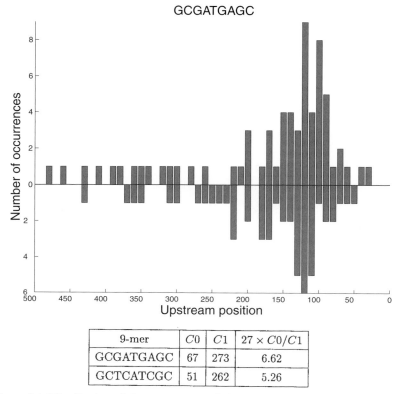

9-mer	$C0$	$C1$	$27 \times C0/C1$
GCGATGAGC	67	273	6.62
GCTCATCGC	51	262	5.26

Figure 6.4. Distribution of the occurrences of the overrepresented 9-mer GCGAT-GAGC and its reverse complement (mirror histogram) across all 500 bp gene-upstream regions in yeast. $C0$ is the total number of occurrences. $C1$ represents the total number of occurrences of all the $27 = 3 \times 9$ 9-mers that differ in only one position from the 9-mer (background model). Under this model, the 9-mer is over six-fold overrepresented.

fraction of the putative motifs detected by these techniques nowadays are typically found also in the TRANSFAC [41] database, or in the current literature, and most must await future experimental verification. In the meantime, overrepresented motifs can be further studied in terms of their patterns of localization and co-occurrence within, for instance, upstream regions and/or their DNA structure. Non-uniform patterns of localization, for instance, can be indicative of biological function. An example of localization pattern is given in Figure 6.4 for the overrepresented 9-mer GCGATGAGC in yeast. When one looks at the 500bp upstream regions of all the genes in yeast, this 9-mer and its reverse complement GCTCATCGC have roughly symmetric distributions with a noticeable peak 50 to 150 bp upstream from the genes they seem to regulate [39, 44, 45]. As far as DNA structure is concerned, it can be analyzed to some extent by using some of the available DNA physical scales [45, 51, 52] (e.g., bendability, propeller twist) and DNA–protein co-crystal structures available in the Protein Data Bank [53]. Typical overrepresented k-mers that have peculiar structural properties include runs of alternating AT which are identical to their own reverse complement and correspond to highly bent or bendable DNA regions (such as the TATA box) or, at the opposite end of the structural spectrum, runs of As or runs of Ts which tend to be very stiff.

All together, these techniques are helping inferential and other data mining efforts aimed at unraveling the "language" of regulatory regions. A somewhat orthogonal approach described in [54] computes for each motif the mean expression profile over a set of array experiments of all the genes that contain the motif in their transcription control regions. These profiles can be useful for visualizing the relationship between genome sequence and gene expression data, and for characterizing the transcriptional importance of specific sequence motifs.

Detection of gene expression differences, clusters of co-regulated genes, and/or gene regulatory motifs are essential steps towards the more ambitious and long-term goal of inferring regulatory networks on a global scale, or even along more specific subcomponents [55, 56, 57] such as a pathway or a set of co-regulated genes. These are the topics of Chapter 8.

REFERENCES

1. Baldi, P., and Brunak, S. *Bioinformatics: The Machine Learning Approach*, 2nd edn. 2001. MIT Press, Cambridge, MA.
2. Alter, O., Brown, P. O., and Botstein, D. Singular value decomposition for genone-wide expression data processing and modeling. 2000. *Proceedings of the National Academy of Sciences of the USA* 97:10101–10106.
3. Baldi, P., and Hornik, K. Neural networks and principal component analysis: learning from examples without local minima. 1988. *Neural Networks* 2:53–58.
4. Roweis, S. EM algorithm for PCS and SPCA. In M. I. Jordan, M. J. Kearns, and S. A. Solla, editors, *Advances in Neural Information Processing Systems*, vol. 10, pp. 626–632. 1998. MIT Press, Cambridge, MA.
5. Scholkopf, B., Smola, A., and Mueller, K. R. Nonlinear component analysis as a kernal eigenvalue problem. 1998. *Neural Computation* 10:1299–1319.
6. Bishop, C. M. Bayesian PCA. In M. S. Kearns, S. A. Solla, and D. A. Cohn, editors, *Advances in Neural Information Processing Systems*, vol. 11, pp. 382–388. 1999. MIT Press, Cambridge, MA.
7. Hand, D., Mannila, H., and Smyth, P. 2001. MIT Press, Cambridge, MA.
8. Eisen, M. B., Spellman, P. T., Brown, P. O., and Botstein, D. Cluster analysis and display of genome-wide expression patterns. 1998. *Proceedings of the National Academy of Sciences of the USA* 95:14863–14868.
9. Alon, U., Barkai, N., Notterman, D. A., Gish, K., Ybarra, S., Mack, D., and Levine, A. J. Broad patterns of gene expression revealed by clustering analysis of tumor and normal colon tissues probed by oligonucleotide arrays.

1999. *Proceedings of the National Academy of Sciences of the USA* 96:6745–6750.

10. Heyer, L. J., Kruglyak, S., and Yooseph, S. Exploring expression data: identification and analysis of co-expressed genes. 1999. *Genome Research* 9:1106–1115.

11. Tamayo, P., Slonim, D., Mesirov, J., Zhu, Q., Kitareewan, S., Dmitrovsky, E., Lander, E. S., and Golub, T. R. Interpreting patterns of gene expression with self-organizing maps: methods and application to hematopoietic differentiation. 1999. *Proceedings of the National Academy of Sciences of the USA* 96:2907–2912.

12. Brown, M. P. S., Grundy, W. N., Lin, D., Cristianini, N., Walsh Sugnet, C., Ares, M. Jr., Furey, T. S., and Haussler, D. Knowledge-based analysis of microarray gene expression data by using support vector machines. 2000. *Proceedings of the National Academy of Sciences of the USA* 97:262–267.

13. Sharan, R., and Shamir, R. CLICK: a clustering algorithm with applications to gene expression analysis. In *Proceedings of the 2000 Conference on Intelligent Systems for Molecular Biology (ISMB00), La Jolla, CA*, pp. 307–316. 2000. AAAI Press, Menlo Park, CA.

14. Cheng, Y., and Church, G. M. Biclustering of expression data. In *Proceedings of the 2000 Conference on Intelligent Systems for Molecular Biology (ISMB00), La Jolla, CA*, pp. 93–103. 2000. AAAI Press, Menlo Park, CA.

15. Xing, E. P., Jordan, M. I., and Karp, R. M. Feature selection for high-dimensional genomic microarray data.

16. Furey, T. S., Christianini, N., Duffy, N., Bednarski, D. W., Schummer, M., and Haussler, D. Support vector machine classification and validation of cancer tissue samples using microarray expression data. 2000. *Bioinformatics* 16:906–914.

17. Golub, T. R., Slonim, D. K., Tamayo, P., Huard, C., Gaasenbeek, M., Mesirov, J. P., Coller, H., Loh, M. L., Downing, J. R., Caligiuri, M. A., Bloomfield, C. D., and Lander, E. S. Molecular classification of cancer: class discovery and class prediction by gene expression monitoring. 1999. *Science* 286:531–537.

18. Alizadeh, A., Eisen M., *et al.* Distinct types of diffuse large B-cell lymphoma identified by gene expression profiling. 2000. *Nature* 403:503–510.

19. Poustka, A., von Heydebreck, A., Huber, W., and Vingron, M. Identifying splits with clear separation: a new class discovery method for gene expression data. 2001. *Bioinformatics* 17 (Supplement 1):S107–S114.

20. Tamayo, P., Mukherjee, S., Rifkin, R. M., Angelo, M., Reich, M., Lander, E., Mesirov, J., Yeang, C. H., Ramaswamy, S., and Golub, T. Molecular classification of multiple tumor types. 2001. *Bioinformatics* 17 (Supplement 1):S316–S322.

21. Tishby, N., Pereira, F., and Bialek, W. The information bottleneck method. In B. Hajek and R. S. Sreenivas, editors, *Proceedings of the 37th Annual Allerton Conference on Communication, Control, and Computing*, pp. 368–377. 1999. University of Illinois.

22. Tishby, N., and Slonim, N. Data clustering by Markovian relaxation and the information bottleneck method. In T. Leen, T. Dietterich, and V. Tresp, editors, *Neural Information Processing Systems (NIPS 2000)*, vol. 13. 2001. MIT Press, Cambridge, MA.

23. Slonim, N., and Tishby, N. The power of word clustering for text classification. In *Proceedings of the European Colloquium on IR Research, ECIR 2001*. 2001.

24. Blatt, M., Wiseman, S., and Domany, E. Super-paramagnetic clustering of data. 1996. *Physical Review Letters* 76:3251–3254.
25. Buhmann, J., and Kuhnel, H. Vector quantization with complexity costs. 1993. *IEEE Transactions on Information Theory* 39:1133–1145.
26. Eppstein, D. Fast hierarchical clustering and other applications of dynamic closest pairs. In *Proceedings of the 9th ACM-SIAM Symposium on Discrete Algorithms*, pp. 619–628. 1998.
27. Spellman, P. T., Sherlock, G., Zhang, M. Q., Iyer, V. R., Anders, K., Eisen, M. B., Brown, P. O., Botstein, D., and Futcher, B. Comprehensive identification of cell cycle-regulated genes of the yeast *Saccharomyces cerevisiae* by microarray hybridization. 1998. *Molecular Biology of the Cell* 9:3273–3297.
28. Bar-Joseph, Z., Gifford, D. K., and Jaakkola, T. S. Fast optimal leaf ordering for hierarchical clustering. 2001. *Bioinformatics* 17 (Supplement 1):S22–S29.
29. Duda, R. O., and Hart, P. E. *Pattern Classification and Scene Analysis*. 1973. Wiley, New York.
30. Everitt, B. S. *An Introduction to Latent Variable Models*. 1984. Chapman & Hall, London.
31. Titterington, D. M., Smith, A. F. M., and Makov, U. E. *Statistical Analysis of Finite Mixture Distributions*. 1985. Wiley, New York.
32. Dempster, A. P., Laird, N. M., and Rubin, D. B. Maximum likelihood from incomplete data via the EM algorithm. 1977. *Journal of the Royal Statistical Society* B39:1–22.
33. Baldi, P. On the convergence of a clustering algorithm for protein-coding regions in microbial genomes. 2000. *Bioinformatics* 16:367–371.
34. Hampson, S., Baldi, P., Kibler, D., and Sandmeyer, S. Analysis of yeast's ORFs upstream regions by parallel processing, microarrays, and computational methods. In *Proceedings of the 2000 Conference on Intelligent Systems for Molecular Biology (ISMB00), La Jolla, CA*, pp. 190–201. 2000. AAAI Press, Menlo Park, CA.
35. Pevzner, P. A., and Sze, S. Combinatorial approaches to finding subtle signals in DNA sequences. In *Proceedings of the 2000 Conference on Intelligent Systems for Molecular Biology (ISMB00), La Jolla, CA*, pp. 269–278. 2000. AAAI Press, Menlo Park, CA.
36. Pevzner, P. A. *Computational Molecular Biology: An Algorithmic Approach*. 2000. MIT Press, Cambridge, MA.
37. Mauri, G., Pavesi, G., and Pesole, G. An algorithm for finding signals of unknown length in DNA. 2001. *Bioinformatics* 17 (Supplement 1):S207–S214.
38. van Helden, J., del Olmo, M., and Perez-Ortin, J. E. Statistical analysis of yeast genomic downstream sequences reveals putative polyadenylation signals. 2000. *Nucleic Acids Research* 28:1000–1010.
39. Brazma, A., Jonassen, I. J., Vilo, J., and Ukkonen, E. Predicting gene regulatory elements in silico on a genomic scale. 1998. *Genome Research* 8:1202–1215.
40. van Helden, J., Andre, B., and Collado-Vides, J. Extracting regulatory sites from the upstream region of yeast genes by computational analysis of oligonucleotide frequencies. 1998. *Journal of Molecular Biology* 281:827–842.
41. Wingender, E., Chen, X., Fricke, E., Geffers, R., Hehl, R., Liebich, I., Krull, M., Matys, V., Michael, H., Ohnhauser, R., Pruss, M., Schacherer, F., Thiele, S., and Urbach, S. The TRANSFAC system on gene expression regulation. 2001. *Nucleic Acids Research* 29:281–284.

42. Vilo, J., and Brazma, A. Mining for putative regulatory elements in the yeast genome using gene expression data. In *Proceedings of the 2000 Conference on Intelligent Systems for Molecular Biology (ISMB00), La Jolla, CA*, pp. 384–394. 2000. AAAI Press, Menlo Park, CA.

43. Bussemaker, H. J., Li, H., and Siggia, E. D. Building a dictionary for genomes: identification of presumptive regulatory sites by statistical analysis. 2000. *Proceedings of the National Academy of Sciences of the USA* 97:10096–10100.

44. Hughes, J. D., Estep, P. W., Tavazole, S., and Church, G. M. Computational identification of *cis*-regulatory elements associated with groups of functionally related genes in *Saccharomyces cerevisiae*. 2000. *Journal of Molecular Biology* 296:1205–1214.

45. Hampson, S., Kibler, D., and Baldi, P. Distribution patterns of locally over-represented k-mers in non-coding yeast DNA. 2002. *Bioinformatics* 18:513–528.

46. Blanchette, M., and Sinha, S. Separating real motifs from their artifacts. 2001. *Bioinformatics* 17 (Supplement 1):S30–S38.

47. Martinez-Pastor, M. T., Marchler, G., Schuller, C., Marchler-Bauer, A., Ruis, H., and Estruch, F. The *Saccharomyces cerevisiae* zinc finger proteins MSN2p and Msn4p are required for transcriptional induction through the stress-response element (STRE). 1996. *EMBO Journal* 15:2227–2235.

48. Wu, A. L., and Moye-Rowley, W. S. GSH1 which encodes gamma-glutamylcysteine synthetase is a target gene for YAP-1 transcriptional regulation. 1994. *Molecular and Cellular Biology* 14:5832–5839.

49. Fernandes, L., Rodrigues-Pousada, C., and Struhl, K. Yap, a novel family of eight bzip proteins in *Saccharomyces cerevisiae* with distinct biological functions. 1997. *Molecular and Cellular Biology* 17:6982–6993.

50. Coleman, S. T., Epping, E. A., Steggerda, S. M., and Moye-Rowley, W. S. Yap1p activates gene transcription in an oxidant-specific fashion. 1999. *Molecular and Cellular Biology* 19:8302–8313.

51. Baldi, P., and Baisnée, P.-F. Sequence analysis by additive scales: DNA structure for sequences and repeats of all lengths. 2000. *Bioinformatics* 16:865–889.

52. Gorm Pedersen, A., Jensen, L. J., Brunak, S., Staerfeldt, H. H., and Ussery, D. W. A DNA structural atlas for *Escherichia coli*. 2000. *Journal of Molecular Biology* 299:907–930.

53. Steffen, N. R., Murphy, S. D., Tolleri, L., Wesley Hatfield, G., and Lathrop, R. H. DNA sequence and structure: Direct and indirect recognition in protein–DNA binding. In *Proceedings of the 2002 Conference on Intelligent Systems for Molecular Biology (ISMB02)*. (2002). (in press)

54. Brown, P. O., Chiang, D. Y., and Eisen, M. B. Visualizing associations between genome sequences and gene expression data using genome-mean expression profiles. 2001. *Bioinformatics* 17 (Supplement 1):S49–S55.

55. van Someren, E. P., Wessels, L. F. A., and Reinders, M. J. T. Linear modeling of genetic networks from experimental data. In *Proceedings of the 2000 Conference on Intelligent Systems for Molecular Biology (ISMB00), La Jolla, CA*, pp. 355–366. 2000. AAAI Press, Menlo Park, CA.

56. Friedman, N., Linial, M., Nachman, I., and Pe'er, D. Using Bayesian networks to analyze expression data. 2000. *Journal of Computational Biology* 7:601–620.

57. Zien, A., Kuffner, R., Zimmer, R., and Lengauer, T. Analysis of gene expression data with pathway scores. In *Proceedings of the 2000 Conference on Intelligent Systems for Molecular Biology (ISMB00), La Jolla, CA*, pp. 407–417. 2000. AAAI Press, Menlo Park, CA.

7

The design, analysis, and interpretation of gene expression profiling experiments

A long-term goal of systems biology, to be discussed in Chapter 8, is the complete elucidation of the gene regulatory networks of a living organism. Indeed, this has been a Holy Grail of molecular biology for several decades. Today, with the availability of complete genome sequences and new genomic technologies, this goal is within our reach. As a first step, DNA microarrays can be used to produce a comprehensive list of the genes involved in defined regulatory sub-circuits in well-studied model organisms such as *E. coli*. In this chapter we describe the use of DNA microarrays to identify the target genes of regulatory networks in *E. coli* controlled by global regulatory proteins that allow *E. coli* cells to respond to their nutritional and physical environments. We begin by describing the design and analysis of experiments to examine differential gene expression profiles between isogenic strains differing only by the presence or absence of a single global regulatory protein which controls the expression of a gene regulatory circuit (regulon) composed of many operons.

Before we can identify the genes of any given regulatory circuit we need to be able to measure their behaviors with accuracy and confidence under various treatment conditions. However, because of the influences of experimental and biological errors inherent in high-dimensional DNA microarray experiments, discussed in Chapter 4, this is not a simple task. Therefore, much of the material of this chapter is devoted to the application and validation of statistical methods introduced in Chapters 5 and 6 to address these fundamental problems. In particular: we describe methods to locate sources of errors and estimate their magnitudes, including global false positive levels; we use data obtained from our experiments to show that the employment of the Bayesian framework described in Chapter 5 allows the identification of differentially expressed genes with a higher level of confidence with fewer replications; and we demonstrate that consistent inferences can be made from data obtained from different DNA microarray formats such as pre-synthesized nylon filter arrays and *in situ* synthesized Affymetrix GeneChips™. In sum, we employ a model system to show that appropriate applications of statistical methods allow the discovery of genes of complex regulatory circuits with a high level of confidence.

We have chosen to use experiments performed with *E. coli* for the discussions of this chapter both because of our own scientific interests and because this model organism offers several advantages for the evaluation of DNA array technologies and data analysis methods that are applicable to gene expression profiling experiments performed with all organisms. Foremost among these advantages is the fact that 50 years of work with *E. coli* have produced a wealth of information about its operon-specific and global gene regulation patterns. This information provides us with a "gold standard" which makes it possible to evaluate the accuracy of data obtained from DNA array experiments, and to identify data analysis methods that optimize the identification of genes differentially expressed because of biological reasons from false positives (genes that appear to be differentially expressed due to chance occurrences exacerbated by experimental error and biological variance). The knowledge we have gained through these analyses with this well-defined model organism gives us confidence that these methods are equally applicable to other less well-defined systems. Indeed, Jacques Monod is credited with saying that "What is true for *E. coli* is true for elephants, only more so."

In the second half of this chapter, we turn our attention to the identification of more complex gene expression patterns involving groups of genes that behave similarly in time and/or across different treatment conditions. In these cases, we emphasize that robustness can be achieved by averaging out noise either by experiment replication with a limited number of treatment conditions or by measuring many samples across time, genotypes, etc. To illustrate this point we describe two different types of experiments. In the first example, we describe experiments that measure the gene expression patterns of *E. coli* cells grown under three different treatment conditions, in the presence and absence of oxygen, and in the presence and absence of a global regulatory protein for genes of anaerobic metabolism in the absence of oxygen. With these experiments we show that, when only three samples (two treatment conditions) are measured, individual experiment replication is required for accurate clustering of genes with similar regulatory patterns. In the second example, we describe experiments that measure the effects of 188 drugs on the expression patterns of 1376 genes in 60 cancer cell line samples. In this case, where 60 samples are measured under 188 conditions, we demonstrate that direct applications of clustering methods to entire data sets reveal robust gene regulatory patterns in the absence of individual experimental replications.

Experimental design

Many experimental designs and applications of gene expression profiling experiments are possible. However, no matter what the purpose of the experiment, a sufficient number of measurements must be obtained for statistical analysis of the data, either through multiple measurements of homogeneous samples (replication) or multiple sample measurements (e.g., across time or subjects). This is basically because each gene expression profiling experiment results in the measurement of the expression levels of thousands of genes. In such a high-dimensional experiment, many genes will show large changes in expression levels between two experimental conditions without being significant. These false positives arise from chance occurrences caused by uncontrolled biological variance as well as experimental and measurement errors.

Experimental errors include variations in procedures involved in growing and harvesting cultures or obtaining biological samples discussed in Chapter 4. These errors can be minimized by appropriate experimental design and technical procedures. On the other hand, biological variance is more difficult to control. Even when two cultures of organisms with identical genotypes are grown under the same conditions differences in gene expression profiles are detected. In addition to these false positives, differentially expressed genes that may not be related to the experimental conditions will be detected when targets prepared from cells of different genotypes or even different tissues of the same genotype are queried. On top of these sources of errors, the quality of DNA microarray data can vary even further depending upon the type of the array and the array manufacturing methods and quality. With these considerations in mind, we employed the general design described below for the *E. coli* experiments of this chapter.

The purpose of the first experiment is to identify the network of genes that are regulated by the global *E. coli* regulatory protein, leucine-responsive regulatory protein (Lrp). Lrp is a global regulatory protein that affects the expression of multiple genes and operons. In most cases, Lrp activates operons that encode genes for biosynthetic enzymes and represses operons that encode genes for catabolic enzymes. Interestingly, the intermediary metabolite, L-leucine, is required for the binding of Lrp at some of its DNA target sites; however, at other sites L-leucine inhibits DNA binding, and at yet other sites it exerts no effect at all. While the expression level of about 75 genes have been reported to be affected by Lrp under different environmental and nutritional growth conditions, its specific role in the regulation of cellular metabolism remains unclear. It has been suggested that it might function to coordinate cellular metabolism with the nutritional state of the environment by monitoring the levels of free L-leucine in the cell.

In spite of the fact that much remains to be learned about the Lrp regulatory network, its many characterized target genes make it an ideal system for the development and assessment of statistical methods for the identification of differentially expressed genes. As these methods are developed and as a better understanding of the gene network regulated by this important protein emerges, a clearer view of its physiological purpose will surely follow.

The experimental design for our Lrp experiment consists of four independent experiments, each performed in duplicate, diagrammed in Figure 4.1. In Experiment 1, Filters 1 and 2 were hybridized with ^{33}P-labeled, random-hexamer-generated, cDNA fragments complementary to each of three RNA preparations (Lrp$^+$ RNA1–3) obtained from the cells of three individual cultures of strain IH-G2490 (Lrp$^+$) These three ^{33}P-labeled, cDNA preparations were pooled prior to hybridizations. Following phosphorimager analysis, these filters were stripped and hybridized with pooled, ^{33}P-labeled cDNA fragments complementary to each of three RNA preparations (Lrp$^-$ RNA1–3) obtained from strain IH-G2491 (Lrp$^-$). In Experiment 2, these same filters were again stripped and this protocol was repeated with ^{33}P-labeled cDNA fragments complementary to another set of three pooled RNA preparations obtained from strains IH-G2490 (Lrp$^+$ RNA 4–6) and IH-G2491 (Lrp$^-$ RNA 4–6) as described above. Another set of filters (Filter 3 and Filter 4) was used for Experiments 3 and 4 as described for Experiments 1 and 2. This protocol results in duplicate filter data for four experiments performed with four independently prepared cDNA target sets. Thus, since each filter contains duplicate spots for each ORF and duplicate filters are hybridized for each experiment, four measurements for each ORF are obtained from each of four experiments [1].

Identification of differentially expressed genes

To identify genes regulated by Lrp, the four background subtracted and globally normalized (Lrp$^+$) or (Lrp$^-$) measurements for each ORF from the control or experimental filters of each of the four experiments dia-

Table 7.1. *Distribution of genes with lowest* p-*values from a comparison of Lrp$^+$ and Lrp$^-$ (control vs. experimental)* E. coli *strains*

p-value	Number of genes
<0.0001	12
<0.0005	30
<0.001	44
<0.005	134
<0.01	208

grammed in Figure 4.1 were averaged and compared to one another. A *t*-test analysis of this four-by-four comparison was carried out and the genes were ranked in ascending order of the *p*-values for each gene measurement based on the *t*-test distribution (Table 7.1). The 44 genes differentially expressed with a *p*-value less than 0.001 are shown in Table 7.2.

Determination of the source of errors in DNA array experiments

While we can be most confident that these genes exhibiting the lowest *p*-values are differentially expressed between the Lrp$^+$ and Lrp$^-$ strains, we also expect false positives – genes that appear to be differentially expressed by chance occurrences driven by experimental and biological variables – even among these genes. Thus, we need some method of determining genes that exhibit low *p*-values by chance given a large number of individual measurements, and we need some method of determining the biological and experimental error inherent in each gene expression level measurement. For this reason, the experimental strategy described above was designed to include a sufficient number of replicate experiments for statistical estimation of the magnitude of the errors contributed by each of these variables.

To estimate the magnitude of errors contributed by differences between filters or among RNA preparations we can determine the reproducibility of results obtained from different arrays hybridized with the same target preparations, or from different target preparations hybridized to the same arrays, respectively.

Table 7.2. *Genes differentially expressed between Lrp$^+$ and Lrp$^-$ (control vs. experimental)* E. coli *strains with a* p-*value less than 0.001*

Gene name[a]	Control (mean)	Experimental (mean)	Control (SD)	Experimental (SD)	p-value	PPDE	Fold
yecI	3.11E-05	8.48E-05	3.35E-06	7.49E-06	8.62E-06	0.99516	2.73
uvrA	1.28E-03	1.04E-03	1.15E-05	3.37E-05	1.70E-05	0.99386	-1.23
gdhA	9.16E-05	2.73E-04	1.52E-05	2.16E-05	2.18E-05	0.99329	2.98
oppB*	7.51E-05	1.14E-03	2.12E-05	3.79E-04	2.48E-05	0.99298	15.12
b2343	2.82E-05	1.02E-04	3.75E-06	1.92E-05	2.67E-05	0.99280	3.61
artP	6.73E-05	4.23E-04	1.24E-05	1.16E-04	3.60E-05	0.99200	6.28
b1810	1.07E-04	2.32E-04	3.65E-06	3.20E-05	4.47E-05	0.99136	2.17
oppC*	2.01E-04	1.08E-03	2.34E-05	3.61E-04	5.44E-05	0.99074	5.38
gltD*	5.28E-04	2.74E-05	1.28E-04	1.42E-05	5.87E-05	0.99049	-19.27
b1330	1.07E-04	1.58E-04	5.44E-06	9.60E-06	7.16E-05	0.98981	1.47
uup	2.02E-04	1.60E-04	6.65E-06	5.72E-06	7.66E-05	0.98956	-1.26
oppA*	1.62E-03	3.16E-02	7.63E-04	1.03E-02	8.45E-05	0.98920	19.44
malE*	3.56E-04	2.01E-04	2.32E-05	2.17E-05	1.16E-04	0.98793	-1.78
oppD*	8.97E-05	6.55E-04	2.76E-05	2.05E-04	1.16E-04	0.98793	7.30
galP	3.75E-04	2.11E-04	2.25E-05	2.40E-05	1.31E-04	0.98740	-1.78
lysU*	1.81E-04	1.24E-03	7.48E-05	2.78E-04	1.44E-04	0.98697	6.87
hybA	3.53E-04	2.47E-04	2.11E-05	1.50E-05	1.49E-04	0.98682	-1.43
hybC	3.54E-04	2.34E-04	2.20E-05	1.81E-05	1.61E-04	0.98646	-1.51
yhcB	4.25E-05	6.84E-05	2.84E-06	6.28E-06	1.68E-04	0.98625	1.61
yifM_2	1.12E-04	6.74E-05	5.35E-06	7.61E-06	1.81E-04	0.98589	-1.66
ilvG*	4.21E-04	9.15E-04	7.55E-05	6.85E-05	2.54E-04	0.98411	2.17
grxB	5.95E-05	3.38E-04	1.92E-05	1.07E-04	2.92E-04	0.98332	5.68
phoP	8.29E-05	2.10E-04	1.20E-05	4.42E-05	3.16E-04	0.98285	2.54
ydjA	1.10E-04	1.79E-04	1.26E-05	1.06E-05	3.54E-04	0.98216	1.62
ydaA	2.61E-04	4.88E-04	3.53E-05	5.45E-05	3.55E-04	0.98214	1.87
yddG	1.77E-04	3.25E-04	2.52E-05	3.37E-05	3.84E-04	0.98165	1.84
emrA	3.58E-04	2.78E-04	2.43E-05	4.57E-06	3.95E-04	0.98147	-1.29
b1685	3.71E-05	2.64E-04	1.20E-05	1.22E-04	4.13E-04	0.98118	7.10
glpA	1.28E-04	8.01E-05	8.54E-06	9.26E-06	4.71E-04	0.98029	-1.59
manA	8.71E-05	2.40E-04	2.16E-05	4.08E-05	4.80E-04	0.98016	2.75
ybeD	1.13E-04	4.01E-04	1.70E-05	1.48E-04	5.15E-04	0.97967	3.55
cfa	2.89E-04	4.89E-04	2.83E-05	6.08E-05	5.16E-04	0.97966	1.69
b3914	6.23E-05	1.97E-04	8.70E-06	5.46E-05	5.44E-04	0.97928	3.16
ybiK	2.03E-04	2.76E-04	1.50E-05	1.47E-05	5.78E-04	0.97884	1.36
yggB	1.73E-04	4.50E-04	3.57E-05	8.50E-05	6.05E-04	0.97850	2.61
amn	4.31E-04	6.51E-04	4.47E-05	4.72E-05	6.07E-04	0.97848	1.51
b1976	1.30E-04	1.77E-04	1.21E-05	6.38E-06	7.56E-04	0.97677	1.36
speB	1.21E-04	3.56E-05	2.09E-05	1.08E-05	7.73E-04	0.97659	-3.40
hdeA	2.40E-04	8.29E-04	8.46E-05	9.90E-05	8.12E-04	0.97619	3.45
pheA	9.11E-05	3.41E-04	3.78E-05	4.17E-05	8.36E-04	0.97595	3.75
gst	3.44E-06	7.24E-05	4.05E-06	2.41E-05	8.57E-04	0.97574	21.01
proC	1.76E-04	6.04E-05	5.17E-05	6.99E-06	8.89E-04	0.97543	-2.91
sdaC*	1.82E-04	9.71E-05	2.02E-05	1.76E-05	8.96E-04	0.97536	-1.87
SerA*	2.90E-03	6.56E-04	1.14E-03	1.12E-04	1.00E-03	0.97405	-4.41

Note:
[a] Known Lrp regulated genes are identified by an asterisk.

To estimate the reproducibility between different filters, we employed a statistical t-test to compare data from each pair of filters hybridized with ^{33}P-labeled cDNA targets prepared from the same pooled RNA samples. For example, each normalized and background subtracted ORF measurement on Filter 1 (IH-G2490 (Lrp$^+$)) of Experiment 1 was compared to the data on Filter 2 (IH-G2490 (Lrp$^+$)) of Experiment 1 (Figure 4.1). Similar comparisons were made between filters hybridized with the same RNA preparations for Experiments 2, 3, and 4. This procedure was repeated for filter pairs hybridized with RNA preparations from the IH-G2491 (Lrp$^-$) strain. The results of these comparisons for the control (Lrp$^+$) strain are shown in Table 7.3. Based on the t-test distribution, and in the absence of experimental error, 12 of the 2547 genes expressed at a level above background in all experiments should exhibit a p-value less than 0.005 based on chance alone ($0.005 \times 2547 = 12$). That is, at this level of measurement accuracy 12 false positives (genes that appear to be differentially expressed) out of the 2547 genes measured are expected, and experimental errors will increase this number of false positives to even higher levels. The data in Table 7.3 show that, in fact, an average of 43 genes with a p-value less than 0.005 is observed. This demonstrates that experimental errors introduced by differences between filters increase the global false positive level of this experiment about threefold beyond that expected by chance.

The data in Table 7.3 demonstrate that even greater errors are introduced by differences among RNA preparations. In this case, the global false positive level expected from chance is elevated nearly tenfold. Thus, while some error is introduced by differences among filters, the major source of error is derived from differences among RNA preparations.

These results illustrate the need to reduce experimental and biological differences among RNA preparations as much as possible. They also illustrate the advantage of pooling RNA preparations from independent samples for each experiment. For example, the RNA preparations used here were pooled from three independent cultures. However, when single RNA preparations were used, the false positive levels were more than twice the levels reported in Table 7.3.

Estimation of the global false positive level for a DNA array experiment

The above results demonstrate the necessity of determining the global false positive level of a gene expression profiling experiment. The global false positive level reflects all sources of experimental and biological variation inherent in a DNA array experiment. With this information, a global level

Table 7.3. *False positive level elevations contributed by differences among filters and target preparations*

Filter Comparisons	$p<0.0001$	$p<0.0005$	$p<0.001$	$p<0.005$	$p<0.01$	$p<0.05$
Filter 1 vs.2 of Exp. 1	2	3	7	38	59	298
Filter 1 vs.2 of Exp. 2	1	9	15	57	119	478
Filter 3 vs.4 of Exp. 3	0	2	5	31	68	314
Filter 3 vs.4 of Exp. 4	3	4	9	47	96	393
False positive levels observed (avg.)	2	5	9	43	86	371
False positive levels expected	0	1	3	12	25	127
Target Prep. Comparisons						
Filter 1 of Exp. 1 vs. Exp. 2	3	14	31	113	224	687
Filter 2 of Exp. 1 vs. Exp. 2	3	10	21	126	215	716
Filter 3 of Exp. 3 vs. Exp. 4	8	10	31	131	230	642
Filter 4 of Exp. 3 vs. Exp. 4	1	5	12	75	163	551
False positive level observed (avg.)	4	12	24	111	208	648
False positive level expected	0	1	3	12	25	127

of confidence can be calculated for differentially expressed genes measured at any given statistical significance level. For example, consider the case where 10 genes are observed to be differentially expressed with a *p*-value less than 0.001 when control data is compared to control data, and 100 genes are differentially expressed with a *p*-value less than 0.001 when control data is compared to experimental data. In this case, since 10 false positives are expected from this data set it is reasonable to infer that we can be only 90% confident that the differential expression of any given gene in the set of 100 is biologically meaningful. This example demonstrates that although the confidence level based on the measurement for an individual gene may exceed 99.9% for two treatment conditions (local confidence 0.001), the confidence that this gene is differentially expressed might be only 90% (global confidence 0.9). This example defines an *ad hoc* method of comparing control to control data to derive an estimate of an experiment-wide false positive level.

An ad hoc *empirical method*

To illustrate this *ad hoc* method for the estimation of false positive levels, we averaged the four measurements for each gene from the duplicate control filters of each experiment hybridized with labeled targets from the control strain (IH-G2490 (Lrp$^+$)). We compared these averaged values of the control data from Experiences 1 and 3 to the averaged values of Experiments 2 and 4. In another analysis, we compared the control data from Experiments 1 and 4 to the averaged values of Experiments 2 and 3. These particular two-by-two (control vs. control) comparisons were chosen because they average across experimental errors and biological differences both among filters and RNA preparations. A *t*-test analysis of the data from these comparisons was performed and the genes were ranked in ascending order of the *p*-values for each gene measurement based on the *t*-test distribution. The results of this statistical analysis are shown in Table 7.4.

The data in Table 7.4 show that in the control vs. control data no genes exhibit a *p*-value less than 0.0001. However, an examination of the *p*-values observed when the control data are compared to the experimental data (Table 7.1) shows that six genes are differentially expressed with a *p*-value less than 0.0001. Thus, we can be fairly certain that these 12 genes are differentially expressed because of biological reasons and not by chance occurrence. This is the good news. The bad news is that we know more than 12 genes are regulated by Lrp. These results show us that, given the experimental errors inherent in this experiment, the differentially expressed levels of most genes cannot pass a statistical test as stringent as this. Thus, to identify

Table 7.4. *Distribution of genes with lowest* p-*values from a comparison of Lrp$^+$ to Lrp$^+$ (control vs. control)* E. coli *strains*

p-value	Number of genes
<0.0001	0.0
<0.0005	0.25
<0.001	1.00
<0.005	3.75
<0.01	7.25

other differentially expressed genes we must lower the stringency of our statistical criterion. The data in Table 7.1 show that as the p-value is raised to 0.005 we observe an additional 122 genes that are differentially expressed at this threshold level. At the same time, raising the statistical threshold to 0.005 reveals an average of 3.75 genes that appear differentially expressed with a p-value equal to or less than 0.005 when the control data sets are compared to one another (Table 7.4). This means that, given this complete data set from four replicate experiments, we expect at least 3.75 false positives among the 134 genes differentially expressed with a p-value equal to or less than 0.005. Therefore, our global confidence in the identification of any one of these 134 genes as differentially expressed genes is 97%.

It should be emphasized that relaxing the p-value threshold rapidly increases the average number of false positives in the control (Lrp$^+$ vs. Lrp$^+$) data sets relative to the number of genes differentially expressed at the same p-value in the experimental (Lrp$^+$ vs. Lrp$^-$) data set and, therefore, decreases the confidence with which differentially expressed genes can be identified.

A computational method

David Allison and colleagues have described a computational version of this *ad hoc* method for estimating the false positive level [2]. The basic idea is to consider the p-values as a new data set and to build a probabilistic model for this new data. When there is no change (i.e., no differential gene expression) it is easy to see that the p-values ought to have a uniform distribution between 0 and 1. In contrast, when there is change, the distribution

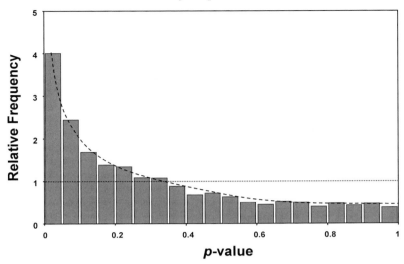

Figure 7.1. Distribution of the *p*-values from the Lrp⁺ vs. Lrp⁻ data. The fitted model (dashed curve) is a mixture of one beta and a uniform distribution (dotted line).

of *p*-values will tend to cluster more closely to 0 than 1, i.e., there will be a subset of differentially expressed genes with "significant" *p*-values (Figure 7.1). One can use a mixture of beta distributions (Chapter 6 and Appendix B) to model the distribution of *p*-values in the form

$$P(p) = \sum_{i=0}^{K} \lambda_i \mathcal{B}(p; r_i, s_i) \tag{7.1}$$

For $i=0$, we use $r_0 = s_0 = 1$ to implement the uniform distribution as a special case of a beta distribution. Thus $K + 1$ is the number of components in the mixture and the mixture coefficients λ_i represent the prior probability of each component. In many cases, two components ($K = 1$) are sufficient but some times additional components are needed. In general, the mixture model can be fit to the *p*-values using the EM algorithm, as in Chapter 6, or other iterative optimization methods to determine the values of λ, r, and s parameters.

From the mixture model, given *n* genes, the estimate of the number of genes for which there is a true difference is $n(1 = \lambda_0)$. Similarly, if we set a threshold *T* below which *p*-values are considered significant and representative of change, we can estimate the rates of false positives and false negatives. For instance, the false positive rate is given by

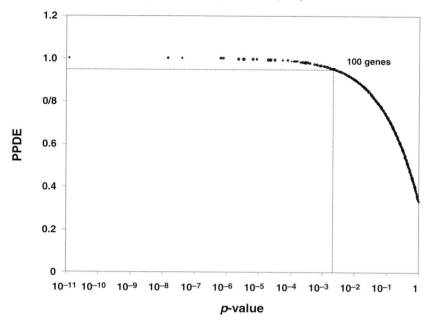

Figure 7.2. Relationship between posterior probability of differential expression (PPDE) and *p*-value. The dotted line correlates the number of genes differentially expressed with a PPDE of 0.95 that are measured with a *p*-value less than 0.0034.

$$\frac{P(p < T \wedge no - change)}{P(p < T)} = \frac{\lambda_0 T}{\lambda_0 T + \sum_{i=1}^{K} \lambda_i \int_0^T \mathcal{B}(p; r_i, s_i) \, dp} \qquad (7.2)$$

The posterior probability for differential expression (PPDE) can then be calculated for each gene in the experiment with *p*-value *p* as

$$\text{PPDE} = P\,(change|p)\ \frac{\sum_{i=1}^{K} \lambda_i \mathcal{B}(p; r_i, s_i)}{\sum_{i=0}^{K} \lambda_i \mathcal{B}(p; r_i, s_i)} = \frac{\sum_{i=1}^{K} \lambda_i \mathcal{B}(p; r_i, s_i)}{\lambda_0 + \sum_{i=1}^{K} \lambda_i \mathcal{B}(p; r_i, s_i)} = \quad (7.3)$$

The distribution of *p*-values from our Lrp$^+$ vs. Lrp$^-$ data is shown in Figure 7.1, and a plot of PPDE values vs. *p*-values is shown in Figure 7.2.

A comparison of the *ad hoc* method for determining the global significance for the differential expression of a given gene and the computational method is presented in Table 7.5. It is satisfying to see that these data compare quite well.

Table 7.5. *Determination of confidence level for differentially expressed*
genes

p-value	Number of genes (control vs. control)	Number of genes (control vs. experimental)	% Confidence (*ad hoc*)	Posterior probability of differential expression
<0.0001	0.00	12	100.00	0.989
<0.0005	0.25	30	99.2	0.980
<0.001	1.00	44	97.7	0.975
<0.005	3.75	134	97.2	0.955
<0.01	7.25	208	96.5	0.944

Improved statistical inference from DNA array data using a Bayesian statistical framework

Although there is no substitute for experimental replication, the confidence in the interpretation of DNA array data with a low number of replicates can be improved by using the Bayesian statistical approach developed in Chapter 5 which employs methods for robust estimate of variance. In this approach, the estimate of the variance of the expression level of a gene is improved by including the variance of additional genes in the neighborhood of the gene under consideration. In the implementation employed here, the neighborhood is defined in terms of genes with similar expression levels. More specifically, the variance of any gene within any given treatment can be estimated by the weighted average of the empirical variance of the gene calculated across experimental replicates and a gene-specific background variance estimated by pooling neighboring genes contained within a window of similar expression levels. Here the weight given to the background variance is a decreasing function of the number of experimental replicates. This leads to the desirable property that as additional replications are performed, the plain and regularized t-test converge on a common set of differentially expressed genes [3, 4].

Commonly used software packages for microarray data analysis do not possess algorithms for implementing Bayesian statistical methods. However, our statistical program, CyberT (Chapter 5 and Appendix D) does accommodate this approach. In the experiments described here, we use the statistical tools incorporated into CyberT to compare and analyze data from the gene expression profiling experiment with the *E. coli* IH-G2490 (Lrp$^+$) and IH-G2491 (Lrp$^-$) strains described above.

Because the Lrp$^+$ and Lrp$^-$ gene expression profiling experiment described above was only replicated four times, and because it is known that the *t*-test does not perform well with a small number of replicate samples, we wished to determine if the identification of known differentially expressed genes could be improved by adding the Bayesian statistical framework of Chapter 5 to this statistical analysis. We therefore used the CyberT software to reanalyze the data from this experiment. Parameters values were set at ws = 101 (window size) and $K = 10$, corresponding to a neighborhood of 100 genes, and a total (experimental + background) of 10 points per gene to estimate the variance (see Chapter 5). A comparison of the results of the analysis employing a simple *t*-test and a regularized *t*-test is shown in Table 7.6.

The simple *ad hoc* method of comparing controls to controls can be used to demonstrate that the number of false positives expected at a given *p*-value is lower when the Bayesian statistical framework is employed. For example, only 2 false positives are expected at a *p*-value threshold less than 0.005 with the Bayesian regularization, whereas 3.75 false positives are expected at this same *p*-value threshold with the *t*-test alone. At the same time, 188 genes differentially expressed genes with a *p*-value less than 0.005 are observed with the regularized *t*-test, whereas only 134 genes are identified at this same threshold with the simple *t*-test (Table 7.6). Thus, since more genes are identified with a lower false positive level and a higher probability of differential expression, the confidence placed on these identifications is increased. In other words, complementing the empirical variance of the four experimental measurements for each gene with the corresponding background variance within an experiment improves our confidence in the identification of differentially expressed genes, and the number of genes that can be identified at a given *p*-value threshold base on a *t*-test distribution.

While the data in Table 7.6 show that the Bayesian statistical approach using a regularized *t*-test identifies more genes with a higher level of global confidence than the simple *t*-test, the natural question that arises is whether these genes are true positives. That is, whether these are Lrp-regulated genes. This question is answered by the data shown in Figure 7.3.

Of the 44 genes differentially expressed between Lrp$^+$ and Lrp$^-$ strains with a *p*-value less than 0.001 identified by a simple *t*-test, 10 are known to be Lrp-regulated (Table 7.2). However, among the 39 genes differentially

Table 7.6. *Comparison of DNA array data analyzed with a simple t-test and a regularized t-test*

	t-test			Regularized t-test			
p-value	Number of genes (control vs. control)	Number of genes (control vs. experimental)	% Confidence	p-value	Number of genes (control vs. control)	Number of genes (control vs. experimental)	% confidence
<0.0001	0.0	12	100	<0.0001	0.0	39	100
<0.0005	0.25	30	99.2	<0.0005	0.25	62	99.6
<0.001	1.00	44	97.7	<0.001	0.50	79	99.4
<0.005	3.75	134	97.2	<0.005	2.00	188	98.9
<0.01	7.25	208	96.5	<0.01	3.75	268	98.6

(a)

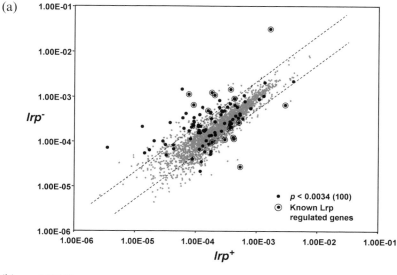

(b)

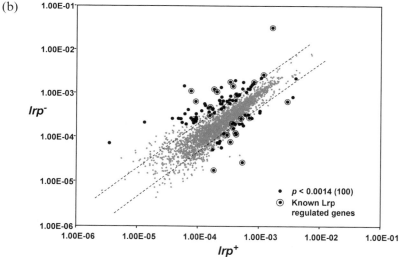

Figure 7.3. Scatter plots showing the mean of the fractional mRNA levels obtained from eight filters hybridized with ^{33}P-labeled cDNA targets prepared from three pooled RNA preparations extracted from *E. coli* K12 strains IH-G2490 (Lrp$^+$) and IH-G2491 (Lrp$^-$). The larger black dots identify 100 genes differentially expressed between strains IH-G2490 and IH-G2491 with (A) *p*-values less than 0.004 and (B) *p*-values less than 0.002 based on a simple *t*-test distribution. The circled black dots identify genes known to be regulated by Lrp. The gray spots represent the relative expression levels of each of the 2885 genes expressed at a level above background in all experiments. The dashed lines demarcate the limits of twofold differences in expression levels.

expressed between Lrp^+ and Lrp^- strains with a p-value less than 0.0001 identified by the Bayesian approach, 17 are known to be Lrp-regulated (Table 7.7). Why does the regularized t-test identify more Lrp-regulated genes? The answer lies in the fact that all the genes identified to be differentially expressed with a p-value less than 0.005 with the regularized t-test exhibit fold changes greater than 1.7-fold (Figure 7.3B). However, many genes identified to be differentially expressed with a p-value less than 0.005 with the simple t-test exhibit fold changes as small as 1.2-fold (Figure 7.3A). Furthermore, the 100 genes with the lowest p-values identified as differentially expressed by both methods contain only 43 genes in common. Thus, many of the genes identified by the simple t-test that are excluded by the Bayesian approach are genes that show small fold changes. In general, these genes with small fold changes identified by the simple t-test are associated with "too good to be true" within-treatment variance estimates, reflecting underestimates of the within-treatment variance when the number of observations is small. The elimination of this source of false positives by the Bayesian approach improves the identification of true positives. However, although this is desired, genes that are truly differentially expressed with small fold changes in the range of 1.2- to 1.7-fold will also be eliminated by the Bayesian approach. For example of the 16 genes of the top 100 with the lowest p-values identified by the simple t-test that are known to be regulated by Lrp, one was not identified by the Bayesian method. This Lrp-regulated gene that did not pass the regularized t-test was the *sdaC* gene, previously reported to be regulated by Lrp and measured to be regulated 1.9-fold in the experiment performed with the DNA arrays [5]. Nevertheless, although this gene is lost, the overall performance of the regularized t-test surpasses that of the simple t-test, and most researchers are interested in discovering genes that are differentially expressed with large fold changes.

At first glance it might appear that the Bayesian approach validates the often-used 2-fold rule for the identification of differentially expressed genes. That is, the identification of genes differentially expressed between two experimental treatments with a fold change greater than 2 in, for example, three out of four experiments. This type of reasoning is based on the intuition that larger observed fold changes can be more confidently interpreted as a stronger response to the experimental treatment than smaller observed fold changes, which of course is not necessarily the case. An implicit assumption of this reasoning is that the variance among replicates within treatments is the same for every gene. In reality, the variance varies among genes (for example, see Figure D.3, Appendix D) and it is

Table 7.7. Genes differentially expressed between Lrp$^+$ and Lrp$^-$ (control vs. experimental) E. coli strains with a p-value less than 0.0001 identified with a regularized t-test

Gene name[a]	Control (mean)	Experimental (mean)	Control (SD)	Experimental (SD)	p-value	Fold	Posterior probability of differential expression
oppA*	1.62E-03	3.16E-02	7.63E-04	1.03E-02	5.14E-13	19.44	1.00000
lysU*	1.81E-04	1.24E-03	7.48E-05	2.78E-04	8.88E-10	6.87	0.99999
oppB*	7.51E-05	1.14E-03	2.12E-05	3.79E-04	1.02E-09	15.12	0.99999
oppC*	2.01E-04	1.08E-03	2.34E-05	3.61E-04	3.26E-09	5.38	0.99998
oppD*	8.97E-05	6.55E-04	2.76E-05	2.05E-04	2.69E-08	7.30	0.99995
serA*	2.90E-03	6.56E-04	1.14E-03	1.12E-04	4.08E-08	-4.41	0.99994
ftn	2.36E-04	1.38E-03	1.29E-04	5.46E-04	2.27E-07	5.84	0.99984
rmf	5.79E-05	1.47E-03	4.68E-05	3.35E-04	2.75E-07	25.43	0.99982
hdeA	2.40E-04	8.29E-04	8.46E-05	9.90E-05	2.99E-07	3.45	0.99982
$ilvP_G$::lacY*[b]	3.68E-04	1.47E-03	4.56E-05	8.10E-04	3.39E-07	3.99	0.99980
hdeB	3.99E-04	1.98E-03	2.58E-04	5.59E-04	4.50E-07	4.96	0.99977
$ilvP_G$::lacA*[b]	3.31E-04	1.83E-03	1.74E-04	7.48E-04	5.64E-07	5.53	0.99974
artP	6.73E-05	4.23E-04	1.24E-05	1.16E-04	1.42E-06	6.28	0.99957
artI	1.26E-04	5.80E-04	3.79E-05	2.80E-04	2.01E-06	4.60	0.99948
gltD*	5.28E-04	2.74E-05	1.28E-04	1.42E-05	2.34E-06	-19.27	0.99943
ilvG_1*	4.21E-04	9.15E-04	7.55E-05	6.85E-05	4.71E-06	2.17	0.99916
livK*	4.16E-04	1.15E-04	1.47E-04	3.22E-05	6.18E-06	-3.61	0.99903
ybeD	1.13E-04	4.01E-04	1.70E-05	1.48E-04	8.55E-06	3.55	0.99884
livH*	4.05E-04	1.24E-04	8.18E-05	5.50E-05	9.07E-06	-3.26	0.99880
uspA	5.42E-04	1.80E-03	3.07E-04	7.43E-04	9.99E-06	3.32	0.99874

pheA	9.11E-05	3.41E-04	3.78E-05	4.17E-05	1.47E-05	3.75	0.99844
grxB	5.95E-05	3.38E-04	1.92E-05	1.07E-04	1.61E-05	5.68	0.99836
b2253	4.24E-04	9.00E-04	8.15E-05	1.50E-04	1.77E-05	2.12	0.99827
hdhA	1.30E-05	2.14E-04	1.22E-05	2.75E-05	1.86E-05	16.49	0.99822
gst	3.44E-06	7.24E-05	4.05E-06	2.41E-05	2.32E-05	21.01	0.99800
oppF*	1.57E-04	4.90E-04	2.82E-05	2.32E-04	2.59E-05	3.13	0.99787
rpoE	1.71E-04	4.35E-04	4.80E-05	7.29E-05	2.67E-05	2.55	0.99784
yhjE	5.44E-04	1.82E-04	6.88E-05	1.20E-04	2.91E-05	-2.98	0.99773
yggB	1.73E-04	4.50E-04	3.57E-05	8.50E-05	2.91E-05	2.61	0.99773
rpoS	3.35E-04	8.77E-04	1.21E-04	3.07E-04	3.03E-05	2.62	0.99768
b1685	3.71E-05	2.64E-04	1.20E-05	1.22E-04	3.66E-05	7.10	0.99743
livM*	6.80E-04	2.74E-04	1.38E-04	1.55E-04	4.24E-05	-2.48	0.99721
rseA	2.41E-04	5.82E-04	5.61E-05	1.32E-04	4.51E-05	2.42	0.99712
$ilvP_G::lacZ*_b$	8.10E-04	1.81E-03	4.51E-05	6.17E-04	4.60E-05	2.24	0.99709
gdhA	9.16E-05	2.73E-04	1.52E-05	2.16E-05	5.44E-05	2.98	0.99681
livJ*	1.16E-03	2.69E-03	5.03E-04	4.42E-04	5.80E-05	2.32	0.99669
fimA*	3.35E-04	7.82E-05	1.46E-04	3.08E-05	6.35E-05	-4.29	0.99652
trxA	9.05E-05	2.84E-04	2.99E-05	4.29E-05	7.43E-05	3.13	0.99621
ydaR	5.15E-05	2.62E-04	2.61E-05	5.61E-05	8.40E-05	5.08	0.99595

Notes:

[a] Known Lrp-regulated genes are identified with an asterisk.

[b] lac genes under the control of the Lrp-regulated $ilvP_G$ promotor–regulatory region.

critical to incorporate this information into a statistical test. This is made clear by simply examining the scatter plots in Figure 7.3. Here many genes that appear differentially expressed greater than 2-fold do not exhibit p-values less than 0.005 and a global confidence level of at least 95%. This does not mean that these might not be regulated genes, it simply means that they are false negatives that cannot be identified at this level of confidence.

The power and usefulness of the statistical methods described here is that all of the genes examined in a DNA array experiment can be sorted by their p-values and global confidence (PPDE) levels based on the accuracy and reproducibility of each gene measurement. At this point, the researcher can set his/her own threshold level for genes worthy of further experimentation. After all, it is best to know the odds when placing a bet.

The Bayesian approach allows the identification of more true positives with fewer replicates.

As additional replications of DNA array experiments are performed, a simple t-test analysis results in the identification of a more consistent set of up- or down-regulated genes. This is of course because better estimates of the standard deviation for each gene are obtained as the number of experimental replications becomes larger. We have shown above that the use of a Bayesian prior estimate of the standard deviation for each gene used in the t-test further improves our ability to identify differentially expressed genes with a higher level of confidence than the simple t-test. This suggests that a more consistent set of differentially expressed genes identified with a higher level of confidence might be identified with fewer replications when the regularized t-test is employed. Long *et al.* [3] have demonstrated that this is indeed the case. In this work, they used CyberT to compare and analyze the gene expression profiles obtained from a wild-type strain of *E. coli* and an otherwise isogenic strain lacking the gene for the global regulatory protein, integration host factor (IHF), reported by Arfin *et al.* [6]. IHF is a DNA architectural protein that is important for the compaction of the bacterial nucleoid. However, unlike other architectural DNA proteins that bind to DNA with low specificity, IHF also binds to high-affinity sites to modulate expression levels of many genes during transitions from one growth condition to another by its effects on local DNA topology and structure.

Long *et al.* [3] defined genes whose differential expression is likely to be due to a direct effect of IHF, and therefore true positives, as those genes that possess a documented or predicted high-affinity IHF binding site within 500 base pairs upstream of the ORF for each gene, or operon containing the gene. Of the 120 genes differentially expressed between IHF^+ and IHF^-

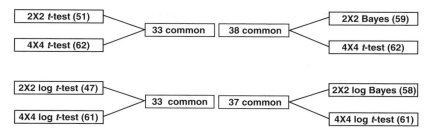

Figure 7.4. Analysis of IHF data with and without Bayesian "treatment". The numbers in parentheses represent the number of genes of the 120 genes with the lowest *p*-values that contain a documented or predicted IHF binding site less than 500 base pairs upstream of each ORF. In each case the raw or log-transformed data from two (2×2) or four (4×4) independent experiments were analyzed either with a simple *t*-test (*t*-test) or regularized *t*-test (Bayes).

strains with the lowest *p*-values identified by a simple *t*-test based on using the data from two independent experiments (2×2 *t*-test), 51 genes containing an upstream IHF site were observed; whereas, a regularized *t*-test (2×2 Bayesian) using these same two data sets identified 59 genes, or 15% more genes with upstream IHF sites (Figure 7.4). Furthermore, comparison of the differentially expressed genes identified by the simple *t*-test or the regularized *t*-test to the differentially expressed genes identified by a simple *t*-test performed on four experimental data sets (4×4 *t*-test) showed that the regularized *t*-test again identifies 15% more genes (38 vs. 33) in common with the genes identified with the 4×4 simple *t*-test. These data demonstrate that replicating an experiment twice and performing a Bayesian analysis is comparable in inference to replicating an experiment four times and using a traditional *t*-test. As data from more replicate experiments are included, the differentially expressed genes identified by the regularized and simple *t*-test analyses converge on a common set of differentially expressed genes.

Deciding a cutoff level for differentially expressed genes worthy of further experimentation

Once we have sorted our gene list based on the PPDE values, we are faced with the problem of deciding a cut-off level for genes worthy of further experimentation. While this is, of course, an empirical decision, common sense criteria can be used. For example, if we observe a gene with a PPDE value of 0.95 that we know, or strongly suspect, is differentially expressed under the experimental treatment conditions employed, then it is reasonable to assume that all genes expressed at a PPDE value greater than 0.95 are differentially expressed genes. It must be kept in mind, however, that even at this high level of confidence we still expect five false positives for each 100 genes in our list and we don't know which five genes these are. Nevertheless, we are safe in concluding that the genes of this list are indeed worthy candidates for further experimentation. On the other hand, a gene that we suspect to be differentially expressed might have a relatively low PPDE value of for example 0.75. While the probability that this gene is differentially expressed is not as high as the genes in our list with PPDE values greater than 0.95, it is still a reasonable candidate for experimental verification. After all, the odds are 3 out of 4 in our favor. Finally, it is quite likely that we will find some genes that we know, or strongly suspect, should be differentially expressed that have low PPDE values. Such false negatives cannot be avoided. Simply because of small fold changes and/or experimental and biological variance, many differentially expressed genes will not be measured with sufficient accuracy for identification. These false negatives must be accepted as "the nature of the beast". In high-dimensional DNA microarray experiments it is just not possible to identify all differentially expressed genes at a high confidence level.

To illustrate these common sense methods of identifying differentially expressed genes we consider the results of our Lrp^+ vs. Lrp^- experiments. In this experiment using a regularized t-test that compares differential gene expression between two otherwise isogenic *E. coli* strains either containing or missing the gene for the global regulatory protein Lrp, we observe 252 genes with a PPDE value greater than 0.95 and 100 with a PPDE value greater than 0.98. Of these 100 genes, 22 are known to be regulated by Lrp. Since we expect only two false positives among these 100 genes we are confident that at least 76 of these genes are newly recognized genes that are differentially expressed between these two *E. coli* strains. These 100 genes are distributed into six functional groups (Figure 7.5). Since the physiological purpose of Lrp is presumed to be the co-ordination of gene expression levels with the nutritional and environmental conditions of the cell, it is pleasing to discover that most of these Lrp-regulated genes encode gene products involved in small and macromolecule synthesis and degradation, as well as gene systems involved in small molecule transport and environmental stress responses.

It is indicative of the power of gene expression profiling experiments that two-thirds of these genes measured with a 98% global confidence were previously unknown to be members of the Lrp regulatory network. Furthermore, nearly one-third of these genes are genes of unknown func-

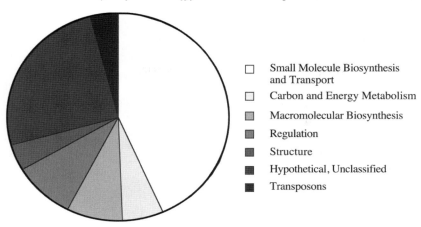

Figure 7.5. Distribution of functions for genes expressed between Lrp$^+$ and Lrp$^-$ *E. coli* strains.

tion. As more experiments of this type are performed and as more functions are assigned to the gene products of these hypothetical ORFs and bioinformatics methods to identify degenerate protein binding sites typical of proteins that bind to many DNA sites are developed, a clearer picture of genetic regulatory networks and their interactions in *E. coli* will emerge. These will be the first steps towards the systems biology goals of Chapter 8.

Nylon filter data vs. Affymetrix GeneChip™ data

The analyses of gene expression profiling experimental data described above clearly demonstrates that experimental errors arise from differences among DNA arrays and the experimental manipulations required for target preparation. When different array formats are used that require different target preparation methods, the magnitudes and sources of these experimental errors are surely different. This raises the question of whether or not results obtained from experiments performed with different DNA array formats can be compared to one another, or indeed if comparable results can be obtained. To address this question, we have assessed the results obtained from DNA array experiments performed with pre-synthesized nylon filters hybridized with ^{33}P-labeled cDNA targets prepared with random hexamer primers from total RNA and *in situ* synthesized Affymetrix GeneChips™

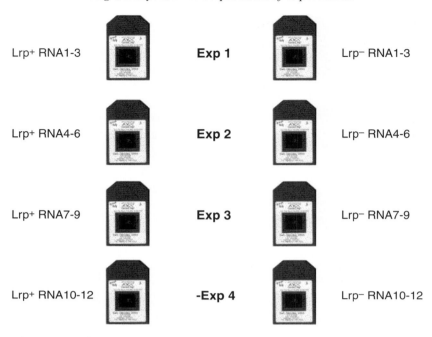

Figure 7.6. Design of the Affymetrix GeneChip™ experiments. See text for details.

hybridized with biotin-labeled targets obtained from the same total RNA preparation.

For the GeneChip™ experiments, the exact same four control and experimental pairs of pooled RNA preparations used in the Lrp$^+$ vs. Lrp$^-$ filter experiments described above were used for hybridization to four pairs of *E. coli* Affymetrix GeneChips™. However, in this case each experiment was not performed in duplicate, and only one measurement for each gene was obtained on each chip. Thus, instead of having four measurements for each gene expression level for each experiment (Figure 4.1), only one measurement was obtained from each GeneChip™ (Figure 7.6). On the other hand, this single measurement is the average of the difference between hybridization signals from 15 perfect match (PM) and mismatch (MM) probe pairs.[1] While these are not equivalent to duplicate measurements because different probes are used, these data do increase the reliability of each gene expression level measurement.

Because each filter experiment was performed in duplicate and each filter contained duplicate probes for each target, it was possible to assess the

---

[1] While the number of perfect match and mismatch probe pairs for the vast majority of *E. coli* ORFs is 15, this number can range from 2 to 20 depending on the length of the ORF.

Table 7.8. *Differential gene expression data for Affymetrix GeneChip™ experiments using CyberT with a regularized* t-test[a]

p-values	Number of genes (control vs. control)[b]	Number of genes (control vs. experimental)	% Confidence (*ad hoc*)
<0.0001	1.00	19	94.7
<0.0005	2.33	27	91.4
<0.001	3.67	32	88.5
<0.005	16.50	53	68.9
<0.01	32.17	66	51.3
<0.05	151.67	152	0.2

Notes:
[a] Calculated with 2128 control and experimental gene expression measurements (AD values from *.CEL file with negative values converted to 0) containing four non-zero values for four experiments.
[b] Calculated by averaging the control or experimental measurements and comparing experiments 1 and 3 vs. 2 and 4 or 1 and 4 vs. 2 and 3 that average data across chips and RNA preparations.

magnitude of errors contributed by differences among the filters and differences among the target preparations (Table 7.3). However, since only one measurement from one GeneChip™ was obtained for each genotype for each experiment, it was not possible to distinguish sources of error contributed by differences among GeneChips™ from differences among target preparations as we did with the filter data. Nevertheless, it was possible to use the *ad hoc* control vs. control method and the computational method of Allison *et al.* [2] and CyberT software to compare data among the four control GeneChips™ hybridized with independent biotin-labeled mRNA targets from *E. coli* strain IH-G2490 (Lrp$^+$). These methods were used to estimate the number of false positives expected at given *p*-value thresholds. These results for the GeneChip™ data, as well as the nylon filter data, are presented in Tables 7.8 and 7.9, respectively

It is clear from these results that the filter data identifies more differentially expressed genes with lower *p*-values and higher confidence levels than the GeneChip™ data. This is not surprising since, as explained above, each gene measurement level in the filter data set is the average of four duplicate measurements from two separate filters, while each gene measurement in the GeneChip™ data set is based on a single measurement from each experiment. In fact, when the top 100 genes with the lowest *p*-values from the nylon filter and GeneChip™ experiments are compared, only 16 genes are common to both lists. This lack of correspondence is likely due to the greater variance among the GeneChip™ measurements and the fact that

Table 7.9. *Differential gene expression data for nylon filter experiments using CyberT with a regularized* t-*test*[a]

p-values	Control vs. control[b]	Control vs. experimental	% Confidence (ad hoc)	PPDE
<0.0005	0.25	62	99.6	0.089
<0.001	0.50	79	99.4	0.085
<0.005	2.00	188	98.9	0.964
<0.010	3.75	268	98.6	0.947
<0.050	17.75	613	97.1	0.878

Notes:
[a]Calculated with 2775 control and experimental gene measurements containing four non-zero values for four experiments.
[b]Calculated by averaging the control measurements of experiments 1 and 3 vs. 2 and 4 or 1 and 4 vs. 2 and 3 that average data across filters and RNA preparations.

fundamentally different DNA array formats are compared. However, when a Bayesian statistical framework (see above and Chapter 5) is applied to the analysis of each data set, the correspondence is doubled and 29 genes are found to be common to both lists. These results further strengthen the conclusions of Long *et al.* [3] that statistical analyses performed with a Bayesian prior identify genes that are up- or down-regulated more reliably than approaches based only on the *t*-test when only a few experimental replications are possible.

The GeneChip™ results described above were obtained from raw data that was background subtracted and normalized to the total signal on each DNA array, and analyzed with the CyberT statistical software. However, Affymetrix has developed its own empirical algorithms for the analysis of GeneChip™ data that is commercially available in a software package called Microarray Suite. Below we compare the identification of differentially expressed genes identified with the CyberT and Microarray Suite software v4.0.

The identification of differentially expressed genes by Microarray Suite is based on genes that fulfill two criteria: (1) the difference between the perfect match and the mismatch values for a given gene probe set must be greater than a defined threshold (CT);[2] (2) the difference between the perfect match

---

[2] CT is a value equal to a standard statistical difference threshold (SDT) of both control and experimental data. SDT is calculated by the relationship $SDT = Q \times M$ where Q is the noise defined as the standard deviation of the pixel intensities of the probe cells exhibiting the lowest 2% intensity values in each of 16 sectors of each GeneChip™, and M is set by default to 2 if a strepavidin–phycoerythrin staining method is used, or 4 if an antibody amplification staining method is used.

Table 7.10. *Number of differentially expressed genes identified by Affymetrix Microarray Suite software*

Number of replicates	Number of differentially expressed genes
1	416–682
2	118–184
3	68–95
4	55

and the mismatch values of the control and experimental chip divided by the perfect match and the mismatch values of the control chip must be greater than the percent change threshold divided by 100, where the percent change threshold is defined by the user (default 80). Based on these criteria the software identifies (calls) differentially expressed genes as marginally increased or decreased, increased or decreased, or not changed.

Since the Affymetrix software allows the comparison of only one GeneChip™ pair at a time, it was run on each of the four independent experiments comparing Lrp genotypes. Each comparison identified between 500–700 genes as marginally increased or decreased, or increased or decreased (Table 7.10). However, filtering identified only 55 genes that were called differentially expressed in all four experiments. Comparison of these 55 genes to the 55 genes exhibiting the lowest *p*-values identified by the CyberT software employing a Bayesian statistical framework revealed 35 genes in common with both lists. Among these, were 17 known Lrp-regulated genes.

These results illustrate several important points. First, they stress the importance of replication when only two conditions are compared. Little can be learned about those genes regulated by Lrp from the analysis of only one experiment with one GeneChip™ pair since an average of 600 genes were identified as differentially expressed, only 55 of which can be reproduced in four independent experiments. Furthermore, in the absence of statistical analysis it is not possible to determine the confidence level and rank the reliability of any differentially expressed gene measurement identified with the Affymetrix software. This is, of course, important for prioritizing genes to be examined by additional experimental approaches. In spite of these limitations, the comparison of the 55 genes identified by examining the results of four GeneChip™ experiments with the results of four nylon filter experiments reveals that differentially genes can be identified with the

Affymetrix Microarray Suite software when multiple replicate experiments are analyzed.

Modeling probe pair set data from Affymetrix GeneChips™

For the *E. coli* Affymetrix GeneChips™, each gene expression level is calculated as the average difference between target hybridization signals from an average of 15 unique perfect match and mismatch probe pairs (Chapter 4). However, the information on expression level provided by the different probes for the same gene is highly variable. In fact, Li and Wong [7] have shown that the variation due to probe effects is five times larger than the variation among different arrays. To account for these probe effects they have developed a statistical model-based method to account for individual probe-specific effects and for handling outliers and image artifacts, as well as a computer program implementing these methods. Modeling the probe pair data sets obtained from the Affymetrix *.dat or *.cel files significantly increases the identification of genes expressed above background and the suitability of the data set for subsequent statistical analyses.

Application of clustering and visualization methods

As we shall see in Chapter 8, a basic paradigm for understanding regulatory networks at a system level involves our ability to perform perturbation experiments. In the case of the Lrp experiments described in the preceding sections, only one variable was changed. We examined the differential gene expression between two *E. coli* strains cultured under identical conditions that differed only by the presence or absence of the Lrp global regulatory protein. In this case, differential gene regulation patterns can be of only two types, they can go up or they can go down. However, when more than two genotypes are compared and/or when experimental parameters involving more than two temporal treatment conditions are analyzed, data mining methods designed to identify more complex gene expression patterns are needed. In the first set of experiments described here, we use clustering and visualization methods to examine the effects of perturbing two variables, one genetic variable and one environmental variable, in a controlled model system. In the second set of experiments, we use these methods to analyze gene expression patterns resulting from the effects of multiple perturbations involving multiple drug treatments on multiple cancer cell types.

Identification of differential gene expression patterns resulting from two-variable perturbation experiments

The efficient transition of *E. coli* cells between aerobic and anaerobic growth conditions requires the activity of a global regulatory protein, Fnr, for the activation or repression of the expression levels of scores of operons encoding many classes of genes. To identify the global changes and adjustments of gene expression patterns that facilitate a change from aerobic to anaerobic growth conditions and to determine the effects of genotype on these gene expression patterns, we analyze *E. coli* gene expression profiles obtained from cells cultured at steady state growth rates under aerobic and anaerobic growth conditions ($+O_2$ or $-O_2$) and under anaerobic growth conditions in the presence and absence of the global regulatory protein for anaerobic metabolism, Fnr ($-O_2 +$Fnr or $-O_2 -$Fnr), in otherwise isogenic strains [8]. Each of these three experiments was performed as described for the Lrp$^+$ and Lrp$^-$ experiment. That is, for each treatment condition three separate total RNA preparations were pooled for each experiment and four independent experiments were performed in duplicate with Sigma-Genosys nylon filters, each containing duplicate spots for each of the 4290 *E. coli* full-length ORF probes. The data sets from each experiment were analyzed with the CyberT program.

As stated above, only two general regulatory patterns can be observed when only two experimental conditions are compared, for example growth in the presence or absence of oxygen. However, when two treatment conditions are compared, at least eight general regulatory patterns are expected. The data in Figure 7.7 diagram the eight basic regulatory patterns that could be observed among three experiments conducted in the presence and absence of oxygen in a Fnr$^+$ strain and in the absence of oxygen in the Fnr$^-$ strain. For simplicity, only three expression levels for each of these three experimental conditions are assumed, low, medium and high.

An intuitive method to identify genes with these regulatory patterns could be to simply use one of the clustering methods described in Chapter 6 on the entire data set. However, in experiments like the one here where only a limited number of replications or sample measurements are practical, resulting in many genes being measured with low confidence levels that result in many false positives as well as false negatives, such a clustering approach could be misleading, placing many genes in the wrong clusters. To circumvent this problem, the approach described here is based on selecting those genes differentially expressed with high confidence levels for the initial clustering. Once the genes of these regulatory patterns are established, one can "fish" for other genes with similar regulatory patterns with

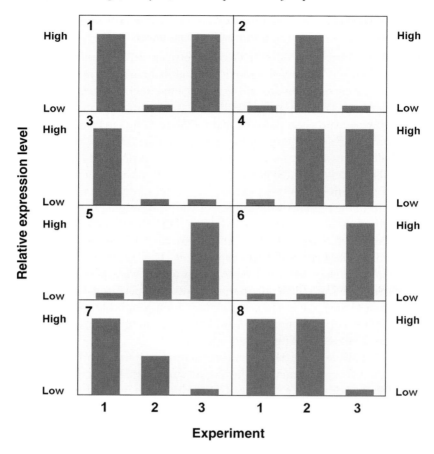

Figure 7.7. Gene expression regulatory patterns expected from the comparison of DNA microarray experiments with one control and two treatment conditions. Control condition (Experiment 1), gene expression levels during growth under aerobic conditions in a Fnr$^+$ *E. coli* strain; Treatment condition 1 (Experiment 2), gene expression levels during growth under anaerobic conditions in a Fnr$^+$ *E. coli* strain; Treatment condition 2 (Experiment 3), gene expression levels during growth under anaerobic conditions in a Fnr$^-$ *E. coli* strain. Each regulatory pattern is designated by a number, 1–8.

lower confidence levels that can be included at the discretion of the investigator.

To identify genes differentially expressed at a high confidence level that correspond to each of the patterns diagrammed in Figure 7.7, the genes differentially expressed due to the treatment condition of Experiments 1 and 2 were sorted in ascending order according to their *p*-values based on the regularized *t*-test as described for the Lrp experiments. Next, the genes differentially expressed due to the treatment condition of Experiments 2 and 3 were sorted in ascending order according to their *p*-values. The 100 genes with the lowest *p*-values present in both lists were selected. These genes exhibited either an increased or decreased expression level between both treatment conditions (i.e., between Experiment 1 and 2, and Experiment 2 and 3). All of these genes possessed *p*-values less than 0.001 representing a 97% global confidence level (PPDE = 0.97) for these gene measurements.

To identify those genes differentially expressed with a high level of confidence under the treatment conditions of Experiments 1 and 2 but expressed at the same or similar levels under the treatment conditions of Experiments 2 and 3, the 500 genes of Experiments 1 and 2 with the lowest *p*-values were compared to the 500 genes with the highest (closest to 1) *p*-values from Experiments 2 and 3. This comparison identified 57 genes that were present in both lists, that is, genes whose regulatory pattern fulfill this criterion. Likewise, to identify those genes differentially expressed under the treatment conditions of Experiments 2 and 3 but expressed at the same or similar levels under the treatment conditions of Experiments 1 and 2, the 500 genes of Experiments 2 and 3 with the lowest *p*-values were compared to 500 genes with the highest *p*-values from Experiments 1 and 2. This comparison identified 48 genes that were present in both lists. These gene lists were combined into a single list of 205 genes differentially expressed under at least one treatment condition.

Hierarchical clustering and principal component analysis

At this point, we have already identified (clustered) those genes differentially expressed with a high level of confidence into the eight regulatory groups of Figure 7.7. In other words, we have developed a "gold standard" for the selection of clustering methods and parameters to be used that are appropriate for this type of an experiment. This is important because all clustering programs will produce clusters, but the members of each cluster will be a function of the clustering parameters chosen. For example, if we use the default settings of the popular GeneSpring software for hierarchical or k-means clustering of our selected gene set we obtain many improperly grouped clusters. Therefore, as discussed in Chapter 6 we found it efficacious to employ a trial-and-error approach to produce clusters of genes exhibiting the regulatory patterns of Figure 7.7. That is, we empirically determined the parameters that result in the hierarchical clustering of our "gold standard" gene set. In other words, we are using our "gold standard" gene set for supervised clustering. The results of this effort are shown in Figure 7.8.

As an independent test, we used PCA to visualize and cluster the same set of 205 genes used for the hierarchical clustering. These PCA clustering results are depicted in Figure 7.9. We were satisfied to see that this independent, unsupervised, method produced results similar to those obtained in Figure 7.8. An advantage of this method is that the PCA results visually highlight correlations among individual gene clusters and identify relationships of outliers that are not as apparent from the hierarchical tree. For example, in the hierarchical tree (Figure 7.8) the single gene of regulatory pattern 5 appears to be most closely related to regulatory patterns 1 and 4. On the other hand, the PCA pattern (Figure 7.9) shows that it is most closely related to regulatory patterns 4 and 6. This discrepancy illustrates a weakness of common hierarchical clustering implementations. As discussed in Chapter 6, each branch of the hierarchical tree lacks orientation and can be freely rotated about its node. In the case here, the branch representing regulatory patterns 1 and 6 can be rotated to bring regulatory pattern 6 next to regulatory pattern 5 as suggested by the PCA analysis. In this respect, it is comforting to know that recent algorithms have been developed to optimize the ordering of the leaves of hierarchical trees (see Chapter 6).

Now that based on our "gold standard" set of differentially expressed genes we are satisfied with our clustering, we can use selected genes of each cluster to "fish" for other genes of the entire data set regulated with a similar pattern that exhibit an arbitrary coefficient of correlation such as, for example, 0.95. To do this we used the "find similar" function of the GeneSpring software package. The number of additional genes to be included in each cluster can be arrived at by selecting a PPDE level cutoff.

Interpretation of clustering results

For the experimental example described here, we can assess the accuracy of our clustering methods of correlating the known regulatory pattern of certain well-studied genes with the patterns identified with the clustering algorithm. However, in other less well-defined systems, such a standard is not available. Nevertheless, if the methods described here, based on a statically reliable subset of genes, produce expected results then it is reasonable to assume that this is a valid method that can be used as a general approach. For example, the genes of patterns 1, 5, and 6 are repressed by Fnr and those of pattern 2, 7, and 8 are activated by Fnr (Figures 7.7 and 7.8). While some of these genes are expected to be affected only indirectly by the presence or

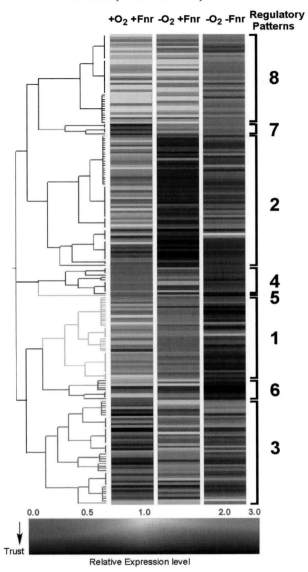

Figure 7.8. Hierarchical clustering of differentially expressed gene regulatory patterns. Experimental cell growth conditions: wild-type *E. coli* K12 strain (Fnr$^+$) under aerobic conditions ($+O_2$ $+$Fnr); wild-type *E. coli* K12 strain (Fnr$^+$) under anaerobic conditions ($-O_2$ $+$Fnr); isogenic *E. coli* K12 strain lacking the Fnr gene ($-O_2$ $-$Fnr) under anaerobic conditions. Each regulatory pattern is identified by numbers that correspond to the regulatory patterns defined in Figure 7.6. The trust parameter is directly related to the mean divided by the standard deviation for each gene measurement.

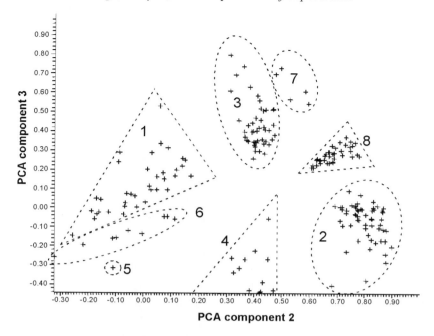

Figure 7.9. Principal component analysis clustering of differentially expressed gene regulatory patterns. A two-dimensional projection onto the plane spanned by the second and third principal components (see Chapter 6). Each cluster is circled by a dotted line. The clusters are numbered according to the regulatory patterns of Figure 7.6.

absence of Fnr, others whose expression is directly regulated by Fnr should possess a DNA binding site(s) upstream of their transcriptional start sites.

Fnr protein is a homodimer that contains symmetrical helix–turn–helix structures that interact with a highly conserved, dyad consensus DNA sequence. This consensus sequence, obtained from mutational analyses and chromosome footprinting experiments performed with nearly a dozen Fnr binding sites, is TTGAT-N4-ATCAA. When Fnr acts as an activator of gene expression it most often binds to a site(s) upstream of and including a site centered at an average distance of 41.5 base pairs before the transcriptional start site of the affected gene or operon. When it acts as a repressor of gene expression it binds to other sites often located near the transcriptional start site of the affected gene or operon [9]. Of the 46 genes repressed by Fnr (patterns 1, 5, and 6; Figures 7.7 and 7.8), 20 contain a documented or predicted Fnr binding site at or near the transcriptional start site with less than a 2 base pair mismatch to the 10 base pair Fnr consensus sequence. Of the

63 genes activated by Fnr (patterns 2, 7, and 8; Figures 7.7 and 7.8), 30 contain an upstream documented or predicted Fnr binding site. Furthermore, since the expression levels of the 57 genes of patterns 3 and 4 are not affected by the presence or absence of Fnr, they are not expected to possess binding sites for this regulatory protein. Again, this is the case. None of these genes possesses an Fnr binding site with less than a 2 base pair mismatch to the 10 base pair Fnr consensus sequence. Thus, the statistical and clustering methods described here produce results consistent with biological expectations.

Clustering data sets resulting from multiple-variable perturbation experiments

Unlike for the analysis of data sets with few replicates or sample measurements, for the analysis of data sets with many sample measurements (e.g., 50 or more) biological variance is averaged out, rigorous statistical analyses become increasingly less necessary, and clustering of total gene sets becomes more feasible. This is dramatically illustrated by the work of Scherf *et al.* [10]. In this study an average-linked dendrogram was generated based on the growth inhibition effects of 118 previously characterized drugs on 60 cancer cell lines of known origins; and, in a separate study, a dendrogram was generated based on the effects of these 118 drugs on the expression patterns of 1376 genes for which expression values were obtained for 56 or more of the 60 cell lines. These data were used to create a two-dimensional clustered image map (CIM) to visualize the relationships between drug activity (y-axis, Figure 7.10) and gene expression patterns (x-axis, Figure 7.10). In this CIM, each individual block represents a high positive or negative correlation between a cluster of genes and a cluster of drugs.

The efficacy of this method is illustrated in the inserts of Figure 7.10. For example, it is known that 5-FU, commonly used to treat colorectal and breast cancer, inhibits both RNA processing and thymidylate synthesis. It is also known that the enzyme dihydropyrimidine dehydrogenase (DPYD) is rate limiting for the degradation of 5-FU. Thus, it was satisfying to observe a highly significant negative correlation between *DPYD* expression level and 5-FU potency in the 60 cell lines (Figure 7.10, insert A). In another example, it is known that many acute lymphoblastic leukemia cell lines lack a functional biosynthetic L-asparagine synthetase (ASNS). This dependence is exploited by treating with L-asparaginase which degrades extracellular L-asparagine. Again, Scherf *et al.* [10] observed a significant negative correlation between *ASNS* expression level and L-asparaginase sensitivity in the 60 cell lines (Figure 7.10, insert B).

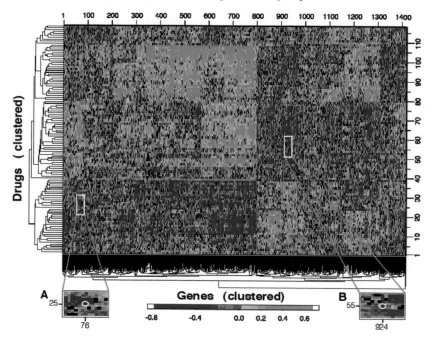

Figure 7.10. Cluster image map (CIM) relating activity patterns of 118 drug compounds to the expression patterns of 1376 genes in 60 cell lines. See text for explanation. (Reproduced with permission from Scherf *et al.*, 2000 [10].)

These results of Scherf and his colleagues [10] clearly demonstrate the usefulness of unsupervised clustering algorithms for the analysis of large data sets. Nevertheless, it should be kept in mind that in most cases where small sets of data are examined unsupervised clustering methods do not perform as well. In these cases, attention must be paid to the experimental error and biological variance inherent in DNA microarray experiments, and statistical methods and supervised clustering procedures of the type described for the Lrp$^+$ vs. Lrp$^-$ experiment described earlier in this chapter should be employed.

REFERENCES

1. She-pin Hung and Hatfield, G. W., manuscript in preparation.
2. Allison, D. B., Gadbury, G. L., Moonseong, H., Fernandez, J. R., Cheol-Koo, L., Prolla, T. A., and Weindruch, R. A mixture model approach for the analysis of microarray gene expression data. 2002. *Computational Statistics and Data Analysis* 39:1–20.

3. Long, A. D., Mangalam, H. J., Chan, B. Y. P., Tolleri, L., Hatfield, G. W., and Baldi, P. Improved statistical inference from DNA microarray data using analysis of variance and a Bayesian statistical framework. 2001. *Journal of Biological Chemistry* 276:19937–19944.

4. Baldi, P., and Long, A. D. A Bayesian framework for the analysis of microarray expression data: regularized *t*-test and statistical inferences of gene changes. 2001. *Bioinformatics* 17(6): 509–519.

5. Calvo, J. M., and Matthews, R. F. The leucine-responsive regulatory protein, a global regulator of metabolism in *Escherichia coli*. 1994. *Microbiology Reviews* 58(3):466–490.

6. Arfin, S. M., Long, A. D., Ito, E. T., Tolleri, L., Riehle, M. M., Paegle, E. S., and Hatfield, G. W. Global gene expression profiling in *Escherichia coli* K12: The effects of integration host factor. 2000. *Journal of Biological Chemistry* 275(38):29672–29684.

7. Li, C., and Wong, W. H. Model-based analysis of oligonucleotide arrays: Expression index computation and outlier detection. 2001. *Proceedings of the National Academy of Sciences of the USA* 98(1): 31–36.

8. Unpublished work from the laboratories of Robert P. Gunsalus and G. Wesley Hatfield.

9. Guest, J. R., Green, J., Irvine, A. S., and Spiro, S. The FNR modulon and FNR-regulated gene expression. In E. C. C. Lin and A. S. Lynch, editors, *Regulation of Gene Expression in* Escherichia coli, pp. 317–342. 1996. Chapman & Hall, New York,

10. Scherf, U., Ross, D. T., Waltham, M., Smith, L. H., Lee, J. K., Tanabe, L., Kohn, K. W., Reinhold, W. C., Myers, T. G., Andrews, D. T., Scudiero, D. A., Eisen, M. B., Sausville, E. A., Pommier, Y., Botstein, D., Brown, P. O., and Weinstein, J. N. A gene expression database for the molecular pharmacology of cancer. 2000. *Nature Genetics* 24(3):236–244.

8

Systems biology

Introduction

In Chapter 7, we studied three global regulatory proteins in *E. coli* (Lrp, IHF, and Fnr). These proteins are responsible for the direct regulation of scores of genes, and through the use of DNA microarrays we were able to establish a fairly comprehensive list of the genes each protein regulates with good confidence. These results have an intuitive graphical representation where nodes represent proteins and directed edges represent direct regulation. Intuitively these simple graphs should capture a portion of the complete "regulatory network" of *E. coli*. Within this network, Lrp, IHF, and Frn are like "hubs", to use an analogy with the well-connected airports of airline flight charts, three of the two dozen or so hubs in the *E. coli* regulatory chart. In spite of their simplicity, these diagrams immediately suggest a battery of questions. How can one represent more complex indirect interactions or interactions involving multiple genes at the same time? Is there any large-scale "structure" in the network associated with, for instance, control hierarchies, or duplicated circuits, or plain feedback and robustness? What is the relationship between the global regulatory proteins (i.e., the hubs) and the less well connected nodes? How are the edges of the hubs distributed with respect to a functional pie chart classification (biosynthesis, catabolism, etc.) of all the genes?

These questions point towards an ever broader set of problems and ultimately whether we can model and understand regulatory and other complex biological processes from the molecular level to the systems level. Such a a systems approach is necessary if we are to integrate the large amounts of data produced by high-throughput technologies into a comprehensive view of the organization and function of the complex mechanisms that sustain life. The dynamic character of these mechanisms and the prevalence of interactions and feedback regulation strategies suggest that they ought to be amenable to systematic mathematical analysis applying some of the methods used in biophysics, biochemistry, developmental biology coupled with more synthetic sciences from chemical engineering, to control theory, and to artificial intelligence and computer science.

Although there are many kinds and levels of biological systems (e.g., immune system, nervous system, ecosystem [1]), the expression "systems biology" [2] is used today mostly to describe attempts at unraveling molecular systems, above the traditional level of single genes and single proteins, focusing on the level of pathways and groups of pathways. A basic methodology for the elucidation of these types of systems has been outlined by Leroy Hood and his group [3].

1. Identify all the players of the system, that is all the components that are involved (e.g., genes, proteins, compartments).
2. Perturb each component through a series of genetic or environmental manipulations and record the global response using high-throughput technologies (e.g. microarrays).
3. Build a global model and generate new testable hypothesis. Return to 2, and in some cases to 1 when missing components are discovered.

Currently, DNA microarrays are one of the key tools in this process, to be used in complement with other tools ranging from proteomic tools such as mass spectrometry and two-hybrid systems for global quantitation of protein expression and interactions, to bioinformatics tools and databases for large-scale storage and simulation of molecular information and interactions. These tools are progressively putting us in the position of someone in charge of reverse-engineering a computer circuit with little knowledge of the overall design but with the capability of measuring multiple voltages simultaneously across the circuit.

Indeed, the long-term goal of DNA microarray technology is to allow us to understand, model, and infer regulatory networks on a global scale, starting from specific networks all the way up to the complete regulatory circuitry of a cell [4, 5, 6, 7, 8, 9, 10, 11, 12].

In this chapter, we first briefly review the molecular players and interactions of biological systems involved in metabolic, signaling, and most of all regulatory networks. The bulk of the chapter focuses on gene regulation and the computational methods that are available to model regulatory networks followed by a brief survey of available software tools and databases. We end the chapter with the large-scale properties and general principles of regulatory networks.

The molecular world: Representation and simulation

It is useful to have a sense of the complexity of the molecular interactions involved in systems biology and the kinds of models that can be attempted today [13]. At the molecular level, biological processes are carried by chemical bonds, both covalent and weak bonds. The formation of bonds always releases some energy. Covalent bonds are the strongest and provide the foundation for the identity of each molecule. They are also used for the transformation of biomolecules and the transfer of biological information, as in the case of phosphorylation and other post-translational modifications of proteins. Weak bonds (mainly hydrophobic bonds, van der Waals bonds, hydrogen bonds, and ionic bonds) are one to two orders of magnitude weaker than covalent bonds and more fleeting by their very nature. The typical energy of a single weak bond is in the 1–7 kcal per mole range (as opposed to 50–100 kcal per mole for covalent bonds) with an average lifetime of a fraction of a second. Collectively, however, large numbers of weak bonds play a fundamental role in the structure of large biomolecules, such as proteins, and in their interactions.

Molecular reactions

While all bonds are based on electrostatic forces and can in principle be derived from first quantum mechanical principles, the detailed simulation of large molecules is already at the limit of current computing resources and molecular dynamics simulations [14]: we still cannot compute the tertiary structure of proteins reliably in the computer, let alone produce detailed simulations of multiple biomolecules and their interactions. Thus, today computational models in systems biology cannot incorporate all the details of molecular interactions but must often operate several levels of abstraction above the detailed molecular level, by simplifying interactions and/or considering pools of interacting species and studying how local concentrations evolve in time.

Consider, for instance, the basic problem of modeling the reaction between an enzyme and a substrate. Substrates are bound to enzymes at active-site clefts from which water is largely excluded when the substrate is bound. The specificity of enzyme–substrate interaction arises mainly from hydrogen bonding and the shape of the active site. Furthermore, the recognition of substrates by enzymes is a dynamic process often accompanied by conformational changes (allosteric interactions).

In a basic model that ignores the details of the interactions we can consider that an enzyme E combines with a substrate S with a rate k_1 to form an enzyme–substrate complex *ES*. The complex in turn can form a product *P* with a rate k_3, or dissociate back into *E* and *S* with a rate k_2

$$E + S \underset{k_2}{\overset{k_1}{\rightleftharpoons}} ES \overset{k_3}{\rightarrow} E + P \qquad (8.1)$$

In the model, the rate of catalysis *V* (number of moles per second) of the product *P* is given by

$$V = \frac{dP}{dt} = V_{max} \frac{[S]}{[S] + K_M} \qquad (8.2)$$

where V_{max} is the maximal rate and $K_M = (k_2 + k_1)/k_3$. For a fixed enzyme concentration [*E*]; *V* is almost linear in [*S*] when [*S*] is small, and $V \approx V_{max}$ when [*S*] is large.

This is the Michaelis–Menten model, which dates back to 1913 and accounts for the kinetic properties of some enzymes *in vitro*, but is limited both in the range of enzymes and the range of reactions it can model *in vivo*. In the end the catalytic activity of many enzymes is regulated *in vivo* in multiple ways including: (1) allosteric interactions; (2) extensive feedback inhibition where, in many cases, the accumulation of a pathway end-product inhibits the enzyme catalyzing the first step of its biosynthesis [15]; (3) reversible covalent modifications such as phosphorylation of serine, threonine, and tyrosine side chains; and (4) peptide-bond cleavage (proteolytic activation). Many enzymes do not obey the Michaelis–Menten formalism and kinetics. An important example is provided by allosteric enzymes where the binding of a substrate to one active site can affect the properties of other active sites in the same protein. The co-operative binding of substrates requires more complex models beyond simple pairwise interactions [16, 17].

The Michaelis–Menten equations also have a somewhat *ad hoc* non-linear form that is resistant to analytical solutions and can only be simulated. To quantitatively implement even such a highly simplified model, one has to know the values of the initial concentrations, as well as the stoichiometric constants that govern the reaction k_1, k_2, and k_3 or at least K_M. The value of K_M varies widely over at least six orders of magnitude and depends also on substrate, and environmental conditions including temperature and pH.

Similar problems arise when we try to model biochemical reactions involved in gene regulation reviewed below. In particular, what is the appropriate level of molecular detail? Clearly such a level is dictated by the system and phenomena one is trying to model, but also by the available data and associated subtle tradeoffs. While molecular data about *in vivo* concentrations and stoichiometric constants remain sparse, emergent high-throughput technologies such as DNA microarrays provide hope that these data will become available. Furthermore, a fundamental property of biological systems is their robustness so that it may not be necessary to know the values of these parameters with high precision. Indeed, models that are a few levels of granularity above the single molecule level can be very successful, as exemplified by the Hodgkin–Huxley model of action potentials in nerve cells, which was derived before any detailed knowledge of molecular ion channels became available. Such higher-level models together with large data sets open the door for machine learning or model fitting approaches, where interactions and other processes are modeled by parameterized classes of generic functions that are capable of approximating any behavior and can be fit to the data. These models are described in detail for gene regulation networks below.

Graphs and pathways

It is also natural to represent molecular interactions using graphs and, for instance, describe a chain of signaling or metabolic reactions as a "pathway". While very useful, however, the notion of a pathway is not always well defined. This is not uncommon in biology where the same can be said of the notion of "gene" or "protein function". To a first approximation a pathway is a set of connected molecular interactions that are part of a biochemical system. The interactions may involve linear processing, but most often they also contain complex branching, convergence, and cycles (e.g., Krebs cycle).

Graphical representations of molecular interactions come with several important caveats one ought to be aware of. To begin with, different graphical representations can be derived depending on whether reactions/interactions are associated with nodes or edges in the graph. Furthermore, in many cases graphical connections represent relations that are more or less indirect and subsume a variety of underlying biological mechanisms and interactions, for instance when "protein A regulates the production of protein B". Thus the graphical representation often hides information such as the presence of intermediary steps or other relevant molecules/metabolites. Multiple molecules may participate in a given reaction and this is not always captured by simple edges in a graph.

Another weakness of a purely graphical representation is that it does not contain relevant temporal or stoichiometric information. Each interaction has its own time-scale and kinetics and these are lost in the graphical representation. Other kinds of relevant information, from DNA/protein

primary sequences, to clinical (dysfunction), and pharmacological properties are also absent. In addition, the cell is a very tightly packed and
crowded environment [18] where in some sense most molecules and pathways stand a chance of interacting with each other, albeit very weakly. At
the system level, one is faced with the issue of cross-talk and its tradeoffs
[19]. In some cases it is advantageous for a molecule to participate in multiple reactions. In other cases, chemical isolation becomes a necessity that is
achieved through chemical specificity or through spatial localization, for
instance by compartmentalization. A two-dimensional graphical representation hides important spatial and geometric considerations such as diffusion, cross-talk, membranes and compartmentalization, active transport,
cell division, and so forth.

Finally, the systems in systems biology arise from complex interactions
between proteins, DNA, RNA, and thousands of smaller molecules.
Although in the end all the systems we are interested in depend on the same
laws of chemistry, it may be useful for orientation purposes to distinguish
three broad subareas within systems biology associated with somewhat
different but also overlapping and mutually constraining networks: metabolic networks, protein networks, and regulatory networks. While our
main topic is regulatory networks, we briefly review the first two for completeness.

Metabolic networks

Metabolic networks represent the enzymatic processes within the cell to
transform food molecules into energy as well as into a number of other
molecules, from simple building blocks such as nucleotides or amino acids,
to complex polymeric assemblies, from DNA to proteins, from which
organelles are assembled. Individual metabolic pathways involve biosynthesis but also biodegradation catalyzed by enzymes.

The major metabolic features of *E. coli*, which are common to all life, are
well studied and well understood (see, for instance, [20]). All the biosynthetic pathways begin with a small group of molecules called the key pre-
cursor metabolites, of which there are 12 in *E. coli*, and from which roughly
100 different building-blocks are derived including amino acids,
nucleotides, and fatty acids. Most biosynthetic reactions require energy and
often involve the breakdown of ATP whereas degradative reactions eventu-
ally generate ATP. *Escherichia coli* has about 4400 genes and on the order of
1000 small metabolites which have fleeting existences between synthetic and
degradative steps. The typical operation along a metabolic pathway is the

combination of a substrate with an enzyme in a biosynthesis or degradation reaction. Typically a metabolite can combine with two different enzymes to be broken down into X or to be used in the biosynthesis of Y. There are however many exceptions to the typical case such as, for example, reactions that are catalyzed by RNA molecules. It is also not uncommon for an enzyme to catalyze multiple reactions and for several enzymes to be needed to catalyze a single reaction.

As we have seen in the discussion of the Michaelis–Menten model, the catalytic activity of enzymes is regulated *in vivo* by multiple processes including allosteric interactions (conformational changes), extensive feedback loops, reversible covalent modifications, and reversible peptide-bond cleavage.

Protein networks

The term "protein networks" is usually meant to describe communication and signaling networks where the basic reaction is not between a protein and a metabolite but rather between two proteins or more. These protein–protein interactions are involved in signal transduction cascades, for instance to transfer information about the environment of the cell captured by a G-protein coupled receptor down to the DNA in the nucleus and the regulation of particular genes. Protein networks can thus be viewed as information processing networks where information is processed and transferred dynamically mainly through protein interactions [21]. Proteins functionally connected by post-translational modifications, allosteric interactions, or other mechanisms into biochemical circuits can perform a variety of simple computations including integration of multiple inputs, coincidence detection, amplification, and information storage with fast switching times, typically in the microsecond range. Symbolically, many of these operations, as well as those involved in regulation (see below), can be represented using the formalisms of Boolean circuits and artificial neural networks. In principle these operations can be combined in circuits to carry virtually any computation.

The resulting circuits have staggering complexity: a typical mammalian call may have on the order of 30000 genes and one billion proteins. Not only do most proteins exist in multiple copies, but also in multiple forms due, for instance, to post-translational modifications. Because the number of genes in a mammal is only one order of magnitude the number of genes in a bacterium, and about two to three times the number of genes in the worm or the fly, biological complexity must arise from the variety of molecular species and the corresponding interactions [22]. A single mammalian cell has on the order of a thousand different channel and receptor proteins on its surface, a hundred G-proteins and second messengers, and a thousand kinases. It has been estimated that up to one-third of all proteins in an eukaryotic cell may be phosphorylated.

Indeed, the concentration and activity of thousands of proteins and other molecules in the cell provide a memory trace of the environment and the entire short-term behavior of the cell, or for that matter even an entire organism, relies on circuits of proteins which receive signals, process them, and produce outputs either in terms of a modification of the internal state of the cell, or a direct impact on the environment, such as mechanical impact in the case of movement (e.g., bacterial chemotaxis) or chemical impact in the case of secretion.

These interactions also can be conceptualized in terms of graphs with linear structures, but also with extensive branching and cycles. The relations between these graphs and the underlying biochemical pathways exist but are not obvious since, for instance, other substrates or metabolites can be present which are not proteins. Issues of cross-talk between protein pathways are important although in general, because of their size and interactions, proteins do not diffuse through the cell as rapidly as small molecules (a molecule of cyclic AMP can move anywhere in a cell in less than one-tenth of a second).

Regulatory networks

Gene expression is regulated at many molecular levels starting from the DNA level, to the mRNA level, to the protein level [23]. How genes are encoded in DNA and their relative spatial location along the chromosomes affect regulation and expression. For instance, in bacteria the genes coding for the chains of heterodimeric proteins are often co-located in the same operon, so that they are expressed at the same time, in the same amounts, and in the same neighborhood. Likewise genes belonging to the same pathway are often co-located and the spatial sequential arrangement of related genes often reflects spatial or temporal patterns of activation. In bacterial chemotaxis, for instance, the spatial arrangement of genes reflects their temporal pattern of activation.

Cascades of metabolic and signaling reactions determine the concentrations of the proteins directly involved in the transcription of a given gene. The proteins interact with themselves and with DNA in complex ways to build transcriptional molecular "processors" that regulate the rate of transcription initiation and transcription termination (attenuation in bacteria).

Transcription is an exceedingly complex molecular mechanism which is regulated at multiple levels that are not completely understood. Further regulation occurs at the level of RNA processing and transport, RNA translation, and post-translational modification of proteins. Increasingly RNA molecules are being found to participate in regulation [24]. On top of these complexities, the degradation of proteins and RNA molecules is also regulated, and new regulatory mechanisms are still being uncovered, such as purge mechanisms of *Salmonella* and *Yersinia* bacteria whereby gene expression in an infected host cell is regulated by secreting regulatory proteins from the pathogen to the host. Since most of these regulatory steps are carried out by other proteins, it is obvious that the regulatory networks of a cell are replete with feedback loops.

Although bacteria are often viewed as simple organisms, their gene regulation is already very complex and exhibits many of the features encountered in higher organisms. Bacterial response to any significant stress, for instance, involves adjustments in the rates of hundreds of proteins besides the specific responders. In bacteria, DNA regulatory regions are relatively short, in the range of a few hundred base pairs, often but not always upstream of the operon being regulated. On average a typical bacterial gene is regulated by less than a handful of transcription factors.

DNA regulatory regions in higher organisms, however, extend over much longer stretches of DNA involving thousands of nucleotides and in many cases regulatory elements are found downstream of the gene being regulated, or even inside introns. For example, very distant 3' regulatory elements in the fly are described in [25, 26]. Often, dozens of transcription factors are involved in the regulation [27, 28] and even a fairly specific function, such as the regulation of blood pressure in humans, is already known to involve hundreds of genes.

In bacteria, the regulation of several genes is fairly well understood and a small number of regulatory networks have been studied extensively, to the point where qualitative and even quantitative modeling is becoming possible. Classical examples are the sporulation choice in *Bacillus subtilis* and the lytic–lysogenic choice in phage lambda [29]. In higher organisms the situation is considerably more complex and detailed modeling studies still focus mostly on the regulation of a single gene and its complexity. The *endo 16* gene in the sea urchin *Strongylocentrus purpuratus* is probably one of the best studied examples still in progress [28, 30, 31] (Figure 8.1). The *cis*-regulatory region of *endo 16* covers over 2000 bp, contains no fewer than seven different interacting modules, with a protein binding map that contains dozens of binding sites and regulatory proteins with a highly

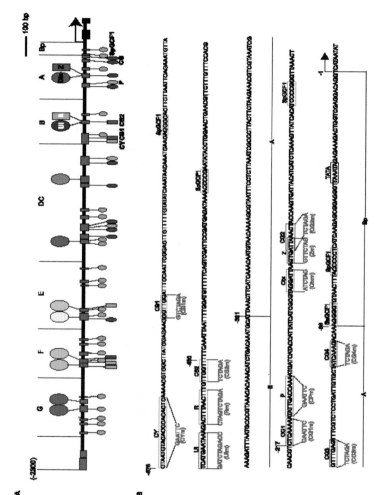

Figure 8.1. Regulatory region of the *endo 16* gene in sea urchin. The region is 2300 bp long and contains distinct functional modules (A to G) in addition to the basal promoter (Bp) region. Factors that bind uniquely in a single region of the sequence are marked above the line representing the DNA. Factors below the line interact in multiple regions. Module G is a general booster for the whole system; F, E, and DC are repressor modules that permit ectopic expression. Module B drives expression later in development, and its major activator is responsible for a sharp, gut specific increase in transcription after gastrulation. Module B interacts strongly with module A which controls the expression of *endo 16* during the early stages of development. (B) DNA sequence of modules B, A, and the basal promoter region. Beneath are target site mutations used in perturbation experiments. (Reprinted with permission from Yuh *et al.*, 2001 and Company of Biologists Ltd.)

non-linear regulatory logic that is complex but amenable to computational modeling.

In gene regulation, there is also a tension between stochastic and deterministic behavior that is not completely understood quantitatively. On a large scale, regulatory mechanisms appear to be deterministic and the short-term behavior of cells and organisms ought to be somewhat predictable. However, at the molecular level, there is plenty of stochasticity in nanoscale regulatory chemistry due to thermal fluctuations and their consequences, such as the random distribution of a small number of transcription factor molecules between daughter cells during the mechanics of cell divisions. In some cases, expression of a gene in a cell can literally depend on a few dozen molecules: a single gene copy and a few transcription factor molecules present in very few copies. Fluctuations and small number of molecules of course influence the exact start and duration of transcription and current evidence suggests that proteins are produced in short random bursts [32, 33]. There are examples, for instance during development, where cells or organisms seem to make use of molecular noise. In most cases, however, regulatory circuits ought to exhibit robustness against perturbations and parameter fluctuations [34, 35, 36]. Quantitative understanding of the role of noise and, for instance, the tradeoffs between robustness and evolvability, are important open questions.

Finally, at its most fundamental level, gene regulation results from a set of molecular interactions that can only be properly understood within a broader context of fundamental molecular processes in the cell, including metabolism, transport, and signal transduction. It should be obvious that signaling, metabolic, and regulatory networks overlap and interact with each other both conceptually and at the molecular level in the cell. Metabolic and signaling networks can be viewed as providing the boundary conditions for the regulatory networks. There are additional important cellular processes, such as transport, that are somewhat beyond these categories and can also play a role in regulation. Furthermore, in the cell, all the processes and their molecular implementation coexist, interact, and also depend on geometrical, structural, and even mechanical constraints and events, such as cellular membranes, compartmentalization, and cell division. In the next sections, we describe the computational tools that are currently available to model regulatory networks.

Computational models of regulatory networks

On the computational side, several mathematical formalisms have been applied to model genetic networks. These range from discrete models, such as Boolean networks [37, 38, 39], to continuous models based on ordinary or partial differential equations, such as continuous recurrent neural network [40] or power-law formalisms [41, 42, 43], to stochastic molecular models, and to probabilistic graphical models and Bayesian networks [44]. Since it is not possible to model all the details of complex molecular interactions, each formalism attempts to simplify certain details and capture the essence of regulatory networks at some level of granularity. The models differ not only by their mathematical structure but also by their level of abstraction. Stochastic molecular models, for instance, are typically closer to the molecular level than the artificial neural network models which in turn are at a lower level of abstraction than Boolean networks or Bayesian networks. The order in which we present the models, however, has more to do with their formal properties and historical use than with their respective levels of abstraction.

A system approach to modeling regulatory networks is essential to understand their dynamics since most of these networks comprise complex interlocking positive and negative feedback loops the effect of which is difficult to capture with our limited intuition. A system approach is essential also to close the loop with experiments: results of model simulations can be compared with the data and discrepancies can be leveraged to suggest new experiments and better models. So far only a few regulatory networks have been studied in enough detail to be amenable to serious modeling. The models that have been produced for these circuits described in the references, however, are already capable of making useful testable predictions either in terms of predicting the behavior under new untested circumstances, or of suggesting errors or missing information in the existing knowledge.

In the past decades limited models of regulatory networks had to be hand-crafted from the literature. Because of the exponential increase in available data, resulting in particular from high-throughput technologies such as DNA microarrays, it is now becoming possible to use more powerful model-fitting techniques to infer the models from the data. Especially when the exact functional form of certain interactions is not known, it is becoming possible to represent them using generic classes of parameterized functions that can approximate any behavior and conduct sensitivity analysis on the parameters.

A system approach to modeling regulatory networks with current computing resources must ignore most of the molecular details of regulation and focus on circuit properties. At this higher level of abstraction, it is reasonable for simplicity to begin with a focus on proteins only. Hence we can imagine a circuit consisting of N proteins where the fundamental step is to model how protein X_i is regulated by all the other proteins in the network. Several high-level models have been proposed to capture these interactions, ranging from discrete Boolean models, to continuous systems of coupled differential equations, to probabilistic models. The list of models we present is by no means meant to be exhaustive but focuses on the best-studied and most successful examples. Overviews of computational models of regulatory networks and their applications can also be found in [10, 11, 13, 45, 46].

Discrete models: Boolean networks

Some of the earliest models of regulatory networks [47, 48] are binary models that assume that a protein or a gene can be in one of two states, active or inactive, on or off, represented by 1 or 0. This binary state varies in time and depends on the state of the other genes and proteins in the network through a discrete equation:

$$X_i(t + 1) = F_i[X_1(t),..., X_N(t)] \tag{8.3}$$

Thus the function F_i is a Boolean function for the update of the ith element as a function of the state of the network at time t. For instance, X_1 is on if and only if X_2 and X_3 are on. In this case, the function F_1 implements a logic AND function. We can also let $\mathbf{X}(t)$ denote the vector state of the system and write $\mathbf{X}(t+1)=[\mathbf{X}(t)]$.

The network is said to be stationary if the functions F_i are fixed and do not vary in time as a function of other external variables and the state of the network itself. The dynamics of a stationary Boolean network is determined by the initial state of the network, the N Boolean functions $F_1,..., F_N$ and how the updates are carried out. States can, for instance, be updated synchronously all at the same time, or sequentially according to a fixed ordering. These are examples of deterministic update schemes that are reproducible exactly. In stochastic update schemes, randomly selected elements or groups of elements in the network are updated at each time step. In general the dynamics of a stationary network is dominated by two kinds of behaviors: convergence to a fixed point or equilibrium state, or convergence to a limit cycle. This is particularly obvious in a deterministic update mode. Because the state space is finite and has only 2^N values, it is clear that any trajectory must either converge to a fixed point for which $\mathbf{X}(t + 1) = \mathbf{X}(t)$ or intersect itself so that $\mathbf{X}(t + \tau) = \mathbf{X}(t)$ for some τ. In the latter case, the trajectory converges to a limit cycle and the smallest value of τ is the period of the cycle. The set of possible starting-points that converge to a given attractor form the basin of attraction for the attractor. States that do not belong to an attractor are called transient states. Current computers can easily manipulate one billion objects. Thus, all of the attractors, as well as their structure, of a Boolean network with on the order of 30 ($2^{30} \approx 10^9$) elements can be computed. The basin of attraction of individual states can be calculated for even larger networks. While these values are encouraging, they are still at least two orders of magnitude below what would be required to simulate large-scale biological regulatory networks, notwithstanding the limitations intrinsic to binary stationary models.

An important parameter of a Boolean network, or any other model of regulatory networks, is its connectivity. The term connectivity refers to the fan-in and fan-out of the connections or interactions of each element in the network. Indeed, the function F_i in Equation 8.3 may depend on $k < N$ inputs only (fan-in k), where k could be quite small with respect to N. There are 2^{2^k} Boolean functions of k variables and $\binom{N}{k}$ choices of k inputs, so that the number of possible Boolean networks with fixed fan-in k is still staggering. It is not entirely clear what are the proper ranges and distributions of fan-in (number of input connections) and fan-out (number of output connections) in biological regulatory networks although, as we have seen, the typical fan-in is likely to be a handful in bacteria and less than a few dozen in higher organisms. In any case, it is reasonable to assume that the fan-in is bounded by a small constant, less than a few dozens. From the study of global regulatory proteins in *E. coli*, it is clear that the fan-out has a broader range and, for the most promiscuous regulatory proteins, exceeds several hundred in bacteria, and could perhaps be as large as a few thousand in metazoan organisms. It is conceivable, however, that at the molecular level the number of interactions is even larger, everything interacting almost with everything else, but with many of these interactions having exceedingly small effects.

An interesting application of Boolean networks has been the study of global properties of large random networks. Random networks can be generated for instance by randomly selecting k inputs for each node and randomly assigning one of the possible Boolean functions, out of the possible 2^{2^k}, to each corresponding interaction. Simulations with N up to $10\,000$ seem to indicate that for low fan-in values k and certain choices of regulatory functions, the random networks tend to exhibit highly ordered dynamics. For example the median number of attractors has been empirically estimated to be about $\sqrt{N}$, with the length of the attractors also proportional to $\sqrt{N}$. Thus for a network with 10 000 elements, this would give about 100 stable points or limit cycles. If we interpret a limit cycle as a major pattern of gene expression associated with a cell type, these simulations predict that the total number of cell types should be in the range of $\sqrt{N}$. This is in good agreement with biological observations corresponding to a few hundred cell types in a vertebrate organism, although this is far from providing a definitive explanation.

While Boolean networks have provided a useful starting-point and enabled partial simulations of large-scale networks, the underlying simplifications of binary states with discrete and possibly synchronous time updates are not very realistic. Thus more complex models must be introduced to capture multiple levels of activity and more complex temporal dynamics. One immediate extension is to consider discrete models where the state of each element is discretized into l discrete levels representing, for instance, l possible concentration levels. The vector of discretized levels at time $t+1$ is then a discrete function of the discretized levels at time t. While the same discretization can be used across all elements, it is possible also to use a discretization scheme that depends on each element. If the variable X_i represents a concentration and if X_i influences m other element in the network, then X_i may have m threshold values $\sigma_1^i < \sigma_2^i < \ldots < \sigma_m^i$. In this case, the levels can be quantized into $m+1$ levels, so that $X_i = j$ if the concentration is between σ_j^i and σ_{j+1}^i, with $X_i = 0$ if the concentration is below σ_1^i and $X_i = m$ if the concentration is above σ_m^i see [49, 50, 51, 52] for details on these generalized logical networks). A more complete treatment of continuous activation levels and continuous time, however, requires the formalisms of differential equations. As we shall see, the discrete multi-level representation can also be combined with differential equations in the qualitative modeling approach described below:

Continuous models: Differential equations

An alternative to the Boolean class of models is obtained when the state variables X are continuous and satisfy a system of differential equations of the form

$$\frac{dX_i}{dt} = F_i[X_1(t),\ldots, X_N(t), I] \tag{8.4}$$

Where the vector $I(t)$ represents some external input into the system. The variables X_i can be interpreted as representing concentrations of proteins, mRNAs, or small metabolites, and systems such as Equation 8.4 have been used to model biochemical reactions including metabolic reactions and gene regulation.

Different rate models can be derived by choosing different approaches for the functions F_i. In most instances, these functions are non-linear. In many cases, the effect of X_j on X_i is mediated by a corresponding level of activation Y_j determined by a regulation function $Y_j = f_j(X_j)$ that is often, but not always, non-linear. In this case, Equation 8.4 can be rewritten as

$$\frac{dX_i}{dt} = H_i[Y_1(t),..., Y_N(t), I] \tag{8.5}$$

It is often reasonable to assume that the function f_i be continuous, differentiable, monotone increasing, and bounded. The simplest such functions are sigmoid or hyperbolic functions but other models are also possible, including discontinuous ones, such as threshold functions or piecewise linear functions. It should be clear that the system could as well be described directly by a system of differential equations on the activations

$$\frac{dY_i}{dt} = G_i[Y_1(t),..., Y_N(t), I] \tag{8.6}$$

and that the roles of concentrations and activations can be switched. The fixed points of a system described by Equation 8.4 are obtained by solving the system of equations $dX_i/dt = 0$ and their stability can be further analyzed by looking at second-order derivatives and the behavior of trajectories in a small neighborhood. If all trajectories tend to converge onto a fixed point, it is a stable attractor. If all trajectories tend to move away from it, it is a repellor. In the mixed case, it is a saddle point. These analyses are relatively straightforward but in general require numerical simulations due to the non-linearity of F_i. The analysis of oscillations and limit cycles is more involved, even at the qualitative level, and typically requires extensive computer simulations.

Discrete delays resulting, for instance, from the time necessary to complete transcription, translation, and diffusion, can be easily introduced in the formalism by writing, in the stationary case,

$$\frac{dX_i}{dt} = F_i[X_1(t - \tau_{i1}),..., X_N(t - \tau_{iN}), I] \tag{8.7}$$

where τ_{ik} is the stationary delay of the effect of variable k on variable i. Models that include multiple, and even a spectrum of, delays for each interaction are also possible. In the few instances where the system in Equation

8.6 can be treated to some extent analytically, the presence of delays usually greatly complicates the analysis. Delays, however, are relatively simple to implement in computer simulations of the system and will not be considered any further here.

Several possible implementations for the functions F_i have been considered in the literature. One possible "neural network" model for the functions F_i is given by the following special case of Equations 8.4 and 8.5:

$$\frac{dX_i}{dt} = -\frac{X_i}{\tau_i} + \sum_j T_{ij} f_j(X_j) + I_i(t) \tag{8.8}$$

The term $-X_i/\tau_i$ represents a decay or degradation term with time constant τ_i and is essential to prevent the system from running away into highly saturated states. Accordingly, the concentration of X_i has a decay component to model degradation, diffusion, and growth dilution at a rate proportional to the concentration itself.

The coefficients T_{ij} are production constants that represent the strengths of the pairwise interactions between the concentrations and can be partitioned into positive excitatory terms and negative inhibitory terms. Alternatively, all the interactions terms could be positive by incorporating the negative signs into interaction-specific activation functions f_{ij} and allowing decreasing activation functions.

The weighted linear average of the activations represents the global influence of the network. The term $I_i(t)$ can be used to incorporate additional external inputs or a noise term. An essentially equivalent formalism is obtained when the non-linearity is applied to the global weighted activation instead:

$$\frac{dX_i}{dt} = -\frac{X_i}{\tau_i} + f_i\left[\sum_j T_{ij}(X_j) + I_i(t)\right] \tag{8.9}$$

Multiple proteins often interact simultaneously to regulate another protein. Formally, such effects can easily be introduced in the form of higher-order interactions terms, for instance of the form

$$\frac{dX_i}{dt} = -\frac{X_i}{\tau_i} + \sum_{jk} T_{ijk} f_j(X_j) f_k(X_k) + \sum_j T_{ij} f_j(X_j) + I_i(t) \tag{8.10}$$

for second-order interactions. Interestingly, equations like 8.8, 8.9, and 8.10 have been used to model neuronal interactions in the artificial neural network literature. Today we know that these artificial neurons are too simple and do not provide good models of the complex dynamics of biological neurons, i.e., the time behavior of the voltage of a neuron is several orders of complexity above the time behavior of the concentration of a typical protein. This is not to say that these artificial neurons do not oversimplify also the behavior of regulatory networks, but the degree of oversimplification may be more acceptable. Thus, artificial neural networks may provide better models of regulatory networks or protein networks [21] than networks of neuronal cells. It should be remarked, however, that both kinds of networks are essentially carrying computations through their dynamical behavior.

Artificial neural network systems, both discrete and continuous, have been extensively studied in the literature. It is well known, for these and other systems of differential equations, that limit cycles require the presence of feedback loops with odd numbers of inhibitory terms [53, 54] (see also [55, 56] for a study of delays and stability), and that feedback loops with even numbers of inhibitory terms tend to drive multiple stationary points. The stability of limit cycles connected with negative feedback loops has been associated with homeostasis, and bifurcation behavior with differentiation processes seen during development as well as rapid switching behavior (e.g., bacterial sporulation).

One important observation is that not all the variables in the systems considered in this section need to be gene product concentrations. In particular, in addition to other molecular species, it is possible to enrich the systems of equations by allowing hidden variables, i.e., considering that the variables X_i can be partitioned into two sets: visible or measurable variables and non-visible variables, as the hidden units of standard artificial networks or latent variables in statistical modeling. Naturally the presence of hidden variables increases the number of parameters and the amount of data required to learn the parameters of the model.

A related and important property of artificial neural network systems is that they also have universal approximation properties [57, 58]. In simple terms, this means that any well-behaved (for instance continuously differentiable) function or system can be approximated to any degree of precision by a neural network, provided one can add hidden state variables (see [59] for a simple proof). These universal approximation results are typically proved for feedforward networks but can be extended to recurrent networks by viewing them as iterations of a forward function. In short, artificial neural networks provide a general formalism that can be used to approximate almost any dynamical system. While universal approximation properties can be reassuring, the central issue becomes learning, i.e., how efficiently we can find a reasonable approximation from a limited set of training data.

Before we discuss the learning issue, it is worth mentioning another very general related formalism developed by Savageau and others [41, 42, 43,

60]. This is the power-law formalism or S-systems introduced in the late 1960s, which can be described by

$$\frac{dX_i}{dt} = \sum_k T_{ik} \prod_j X_j^{g_{ijk}} - \sum_k U_{ik} \prod_j X_j^{h_{ijk}} + I_i(t) \qquad (8.11)$$

where g_{ijk} (resp. h_{ijk}) are kinetic exponents and T_{ik}, U_{ik} are positive rate constants. Thus in Equation 8.11 the terms are separated into an excitatory and an inhibitory (degradation) component. Non-linearity emerges from the exponentiations and the multiplications, which also capture higher-order interactions as in the case of the artificial neural network equations. In fact, in many ways this formalism is similar to the neural network formalism with which it shares the same universal approximation properties as a consequence of Taylor's formula and polynomial approximation. In fact, an S-system can be viewed as a higher-order neural network with exponents in the interactions and where the activation functions f_is are the identity function. By transforming to a logarithmic scale, Equation 8.11 can be converted to a more tractable linear system. As discussed in the references, this formalism has also been applied to model other biochemical pathways, besides regulatory networks, for instance in pharmacodynamics and immunology.

In addition to simulating the networks and understanding the structure of steady states and limit cycles, it is also important to conduct a bifurcation analysis to understand the sensitivity of steady states, limit cycles, and other dynamical properties with respect to parameter values and noise. These issues are also related to the learning problem and how much the available data constrains the model parameters. Thus before proceeding with other classes of models that can be applied to regulatory networks, it is worthwhile considering the learning problem briefly.

Learning or model fitting

We have seen that if we are interested in a reasonably well-behaved regulatory circuit, then we can be assured that there exists an artificial neural network or S-system that can represent it with great fidelity. The question is how to find such a system, and in particular how to learn from it from the data.

For a fixed model structure, we can in principle learn the parameter values by minimization of a mismatch function, such as error, mean square using optimization techniques such as gradient descent for dynamical systems [61]. This optimization process is not always easy, but conceptually it does not pose major problems, provided enough training data is available.

A first-order system containing only pairwise interactions, such as Equation 8.6, has on the order of N^2 parameters which would typically require on the order of N^2 observations, akin to measuring N concentrations at N different time steps. While this number is quite large, the situation may not be as hopeless as it may appear for at least two reasons.

First, precisely because of DNA microarrays and other high-throughput technologies, large numbers of simultaneous measurements are becoming possible. Consider a circuit comprising 20 genes represented by 20 equations. This is a relatively small number of artificial neural network standards. In principle a microarray experiment with 20 time steps could gather an amount of data within one order of magnitude of what is needed to estimate the coefficients T_{ij} of the system. Microarray experiments could be complemented by other high-throughput technology measurements, for instance determining protein levels by a combination of gel and mass spectrometry experiments. Thus while available measurements have been a limiting factor in the past, this situation is changing rapidly. In this context, it is also worth noting that the smallest known genome that can sustain independent life has only about 470 genes [62].

The second reason is that biological systems are both constrained and robust against noise and parameter fluctuations. Because they are constrained, they typically visit only a small fraction of the phase space of possibilities. Because they are robust, small changes in the coefficients should not alter the behavior significantly, at least most of the time. Thus, the model parameters may not need to be determined with great accuracy. In fact these aspects are leveraged in the quantitative modeling approach described in the next section, which ignores exact parameter values and considers, for each species, a finite number of relevant concentration ranges.

Qualitative and even quantitative model-fitting approaches can be attempted today on some of the best-studied regulatory subnetworks, where data is available on most of the players and interactions, through the literature or through complementary high-throughput experiments. On the side of caution, however, one ought to realize that many interactions are not pairwise and this can further increase the complexity of both models and model fitting. In addition, when many of the players are missing, or when the neural network or S-system interaction model is too much of an oversimplification, then many hidden variables become necessary for the models to approximate the data, raising both issues of overfitting the data and being able to interpret the hidden variables in a biologically meaningful way. With hidden nodes, efficient search through the space of possible networks becomes also a central issue. Today we can learn models for the best-studied regulatory switches in bacteria, but developmental circuits in eukaryotes, for instance, are still beyond our reach. Many more examples of regulatory circuits will have to be studied in great detail if we are to push the envelope of what can be modeled.

Qualitative modeling

Lack of data coupled with issues of robustness suggests that it may be useful to pursue a kind of modeling that is more qualitative in nature. One approach to qualitative modeling is to partition the state space into cells delimited by behavioural thresholds and model the behavior of each concentration in each cell by a very simple first-order differential equation with a single rate. Thus, in addition to its robustness, the most attractive feature of qualitative modeling is that it preserves the Boolean logic aspect of regulation while allowing for continuous concentration levels.

More precisely, consider the piecewise linear differential equations

$$\frac{dX_i}{dt} = -\frac{X_i}{\tau_i} + F_i(X_1,..., X_N) \tag{8.12}$$

with

$$F_i(X_1,..., X_N) = \sum_{j \in J(i)} k_{ij} b_{ij}(X_1,..., X_N) \tag{8.13}$$

where k_{ij} are rate parameters and $J(i)$ is a set of indices, possibly empty, which depends on i. The functions b_{ij} take their values in the set $\{0, 1\}$. These functions are the numerical equivalent of Boolean functions and specify the conditions under which a gene is expressed at the rate k_{ij}. More precisely, they are defined in terms of sums and products of step functions [63, 64] s^+ and s^- of the form

$$s^+(X,\sigma) = \begin{cases} 1 : = X > \sigma \\ 0 : = X < \sigma \end{cases} \tag{8.14}$$

with $s^-(X,\sigma) = 1 - s^+(X,\sigma)$ (Figure 8.2). For instance, the equation $dX_3/dt = k_3 s^+(X_1,a)s^-(X_1,b)s^+(X_2,c)s^-(X_2,d)$ expresses the fact that X_3 is expressed or produced at a rate k_3 provided the concentration of X_1 is between a and b and the concentration of X_2 is between c and d (Figure 8.2). More generally, as in the discretized Boolean case, we can associate with each element X_i a bounded sequence of threshold values $0 \le \sigma_1^i, < \sigma_2^i < ... < \sigma_m^i < \max X_i$. These thresholds partition the phase space of all possible states into cells or orthants. The functions b_{ij} take a constant value, 0 or 1, in each orthant. Any combination of orthants can be represented by a

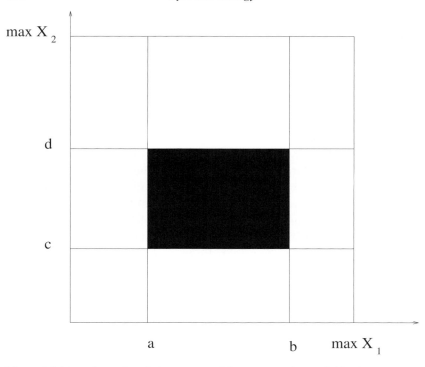

Figure 8.2. Two-dimensional phase space with concentration variables $0 \leq X_1 \leq$ max X_1 and $0 \leq X_2 \leq$ max X_2. The black area corresponds to the region where the function $s^+(X_1, a)\, s^-(X_1,b)s^+(X_2,c)s^-(X_2,d)$ is equal to 1, and where the concentration X_1 is between a and b and the concentration of X_2 is between c and d. In any other cell of the plane, one of the conditions is violated and therefore the function is equal to 0. Any combination of cells in the plane can be described in terms of sums and products of step functions.

generalized Boolean function b_{ij} that is its indicator function, i.e., has a value of 1 on the orthants that belong to the combination. This is easy to see by analogy with the canonical decomposition of a Boolean function into a conjunction of disjunctions with negation. Any number of cells can be described by the sum of the indicator functions associated with each cell. The indicator function of a cell is a product of s^+ and s^- step functions, associated with the conjunction of the corresponding cell threshold conditions.

In each orthant, the total contribution from Equation 8.13 is constant so that the behavior of the system is described by N independent linear equations of the form

$$dX_i/dt = k_i - X_i/\tau_i \qquad (8.15)$$

which have an easy solution $\alpha_i e^{-t/\tau_i} + k_i \tau_i$. Thus within an orthant, all trajectories evolve towards the steady state $X_i = k_i \tau_i$. The steady state of each orthant may lie inside or outside the orthant. If it lies outside, the trajectories will tend towards one or several of the threshold hyperplanes which delimit the orthant. Depending on the exact details of Equation 8.12, these trajectories may be continued in the adjacent orthants or not. Additional important singular steady states may also be located on the threshold planes delimiting the orthants.

Analysis of piecewise linear equations, their behavior and their applications to qualitative modeling of gene regulation and other biological phenomena, can be found in [51, 64, 65, 66, 67, 68, 69, 70, 71, 72, 73, 74].

The global behavior of piecewise linear differential systems can be complex and is not fully understood. However they provide an attractive tool for qualitatively modeling biochemical systems where detailed quantitative information is often very incomplete. In particular the qualitative behavior of the system can be analyzed and represented using a directed graph with one node per qualitative state associated with each orthant. Nodes in the graph are connected whenever the trajectories can pass between the corresponding orthants. This approach has been used in [75] to build a qualitative robust model of the sporulation switch in *B. subtilis* (Figure 8.3). An important observation that is relevant for building or learning such models is that only a very small fraction of the possible cells in phase space is visited by the system during operation. In this particular model, a few dozen states were visited out of several thousand possible states.

Partial differential equations

The models seen so far do not have a spatial structure. Each element in the network is characterized by a single time-dependent concentration level. Many biological processes, however, rely heavily on spatial structure and compartmentalization. During development, for instance, spatial gradients of protein concentrations play an important role. In these cases, it is necessary to model the concentration both in space and time. This leads to reaction–diffusion equations where diffusion terms are added to the reaction component incorporated in the models studied above [76, 77, 78, 79]. This can be done with a discrete formalism or with a continuous formalism using partial differential equations. Reaction–diffusion models of development were initially introduced by Alan Turing [80].

Consider first a discrete linear sequence of K compartment or cells. The concentration $X_i^{(k)}$ of the ith species in cell k depends both on the regulatory effects within cell k and the diffusion effects originated from the two

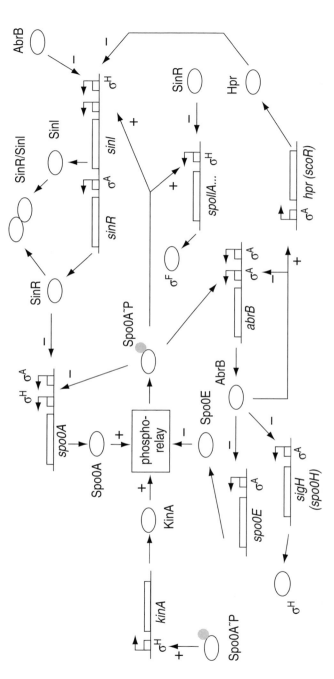

Figure 8.3. Model of genetic regulatory network underlying the initiation of sporulation in *Bacillus subtilis*. Coding region and promoter shown for every gene. Promoters are distinguished by the specific σ factor directing DNA transcription. Signs indicate the type of regulatory interaction (activation/inhibition). The network is centered around a phosphorylation pathway (phosphorelay), slightly simplified in this model, which integrates a variety of environmental, cell cycle, and metabolic signals and phosphorylates Spo0A. The essential function of the phosphorelay is to modulate the phosphate flux as a function of the competing action of kinases and phosphatases (KinA and Spo0E). When a sufficient number of inputs in favor of sporulation accumulate, the concentration of SPo0A~P reaches a threshold value above which it activates several genes that commit the bacterium to sporulation. In order to produce a critical level of Spo0A~P, signals arriving at the phosphorelay need to be amplified and stabilized. This is achieved through several positive and negative feedback loops. Thus the decision to enter sporulation emerges from a complex network. (Reprinted with permission from de Jong *et al.*, 2001.)

neighboring cells in position $k-1$ and $k+1$. The diffusion of products is supposed to be proportional to the concentration differences $X_i^{(k+1)} - X_i^{(k)}$ and $X_i^{(k)} - X_i^{(k+1)}$. This yields the reaction–diffusion model

$$\frac{dX_i^{(k)}}{dt} = F_i^{(k)}(X_1, \ldots, X_N) + D_i^{(k)}(X_i^{(k+1)} - 2X_i^{(k)} + X_i^{(k+1)}) \qquad (8.16)$$

for $1 \leq i \leq N$ and $1 < k < K$. If the system is invariant under translation, for instance in the case of a homogeneous population of cells, then $F_i^{(k)} = F_i$ and $D_i^{(k)} = D_i$. Equation 8.16 is for a one-dimensional system of compartments or cells but it can obviously be generalized to two and three dimensions.

If the number of cells or compartments is large enough over the spatial interval $[0,S]$, we can use a continuous formalism, where the concentrations $X_i(t,s)$ depend both on time t and space s. Assuming translational invariance, the resulting system of partial differential equations can be written as

$$\frac{\partial X_i}{\partial t} = F_i(X_1, \ldots, X_N) + D_i \frac{\partial^2 X_i}{\partial s^2} \qquad (8.17)$$

For $1 \leq i \leq N$ and $0 < s < S$. Assuming that there is no diffusion across the spatial boundaries, the boundary conditions are given by the initial concentrations and $\partial^2 X_i(t,0)/\partial s^2 = \partial^2 X_i(t,S)/\partial s^2 = 0$. Again two- and three-dimensional versions of these equations can easily be derived.

Alan Turing was the first to clearly suggest the use of reaction–diffusion equations to study developmental phenomena and pattern formation. In his original work Turing considered two concentration variables X_1 and X_2, also called morphogens, satisfying Equation 8.16 or 8.17. Even in the case of $N=2$ direct analytical solutions are in general not possible. However if there exists a unique spatially homogeneous steady state X_1^* and X_2^* with $X_1 > 0$ and $X_2 > 0$, the behavior of the system in response to a small perturbation of the steady state can be predicted by simple linearization. In particular, with the boundary conditions above, the deviation ΔX_i from the homogeneous steady state is given by

$$\Delta X_i(s,t) = X_i(s,t) - X_i^* = \sum_{k=0}^{\infty} c_{ik}(t) \cos\left(\frac{\pi k}{S} s\right) \qquad (8.18)$$

for $i = 1, 2$. The functions $\cos(\frac{\pi k}{S} s)$ are the modes or eigenfunctions of the Laplacian operator $\partial^2/\partial s^2$ on $[0,S]$ [76, 81, 82]. The time-dependent coefficients $c_{ik}(t)$ are the mode amplitudes. The steady state is stable with respect to perturbations if all the mode amplitudes decay exponentially in response to a perturbation decomposable into a large number of weighted modes.

When the steady state is unstable, diffusion does not have a homogenizing effect resulting rather into spatially heterogeneous gene expression/concentration patterns. It has been shown that for $N = 2$ this requires one positive feedback loop $\partial F_1 / \partial X_1 > 0$, one negative feedback loop $\partial F_1 / \partial X_2 < 0$, and diffusion constants that favor the rapid diffusion of the inhibitor $D_2 / D_1 > 1$ together with a sufficiently large domain size $S > S_0$ [83]. These activator–inhibitor systems have been used to model the emergence of segmentation patterns during early embryo development in *Drosophila* [77, 78, 84, 85, 86].

The solutions of continuous reaction–diffusion equations $N = 2$ are quite sensitive to initial and boundary conditions, the shape of the domain, and the values of the parameters. Furthermore, there is little experimental evidence that the coupled reaction and diffusion of two morphogens underlies biological pattern formation. The regulatory system that controls embryonic development in *Drosophila* is likely to contain a layered network with hundreds of genes that mutually interact and whose products diffuse through the embryo. More realistic models must use Equation 8.16, for instance, with discrete diffusion and a larger number of interacting species. In [40, 87, 88, 89, 90], a reaction–diffusion network model such as Equation 8.16, with the neural network reaction term of Equation 8.8, is used to model *Drosophila* development.

Although the diffusion component of reaction–diffusion models captures some spatial aspects of regulatory networks, it does not capture all of them. In particular a single set of reaction–diffusion equations can only be used as long as the number of compartments or cells remains fixed. Other events, such as nuclear or cell division, must be modeled at another level. Likewise, different models or additional terms are needed to model spatial organization resulting from processes such as active transport that go beyond diffusion.

Stochastic equations

Ordinary or partial differential equations in general lead to models that are both continuous and deterministic. Both assumptions can be challenged at the molecular level due, as we have seen, to thermal fluctuations and the fact that some reactions may involve only a small number of molecules. Thus, it may be useful to introduce models that are both discrete and stochastic. One possibility is to let X_i represent the number of molecules of species i and to describe the state of the system by a joint probability distribution $P(X_1, \ldots, X_N, t)$ (this can be reformulated in terms of concentration by dividing by a volume factor). If there are M possible different reactions in the system, the time evolution of the probability distribution can be described by a discrete equation of the form

$$P(X_1,\ldots, X_N, t+\Delta t) = P(X_1,\ldots, X_N, t)\left(1 - \sum_{j=1}^{M}\alpha_j\Delta t\right) + \sum_{j=1}^{M}\beta_j\Delta t \quad (8.19)$$

where $\alpha_j\Delta t$ is the probability that reaction j will occur in the interval Δt given that the system is in state $X_1,\ldots, X_N$ at time t, and $\beta_j\Delta t$ the probability that reaction j will bring the system in state $X_1,\ldots, X_N$ in the interval Δt. Letting Δt go to 0 gives the master equation [91]

$$\frac{\partial P}{\partial t} = \sum_{j=1}^{M}(\beta_j - \alpha_j P) \quad (8.20)$$

The master equation describes how the probability distribution, and not the system itself as in the case of ordinary differential equations, evolves in time. In general, analytical solutions are not possible and time-consuming simulations are required. In some cases, a master equation can be approximated by an ordinary stochastic equation, consisting of an ordinary differential equation plus a noise term (Langevin equations [92]).

In the stochastic simulation approach [93, 94], simulations of the stochastic evolution (trajectories) of the system are carried directly rather than simulating the global overarching distribution encapsulated in the master equation. To simulate individual trajectories, the stochastic simulation algorithm needs only to specify the time of occurrence of the next reaction, and its type. This choice can be made at random from a joint probability distribution consistent with Equation 8.19. This is akin to simulating an actual sequence of coin flips rather than studying the averages and variances of a population of runs. This approach is more amenable to computer simulations and has been used in [95, 96, 97] to analyze the interactions controlling the expression of a single prokaryotic gene and to suggest that, at least in some cases, protein production occurs by bursts. Fluctuations in gene expression rates may provide an explanation for observed phenotypic variations in isogenic populations and influence switching behavior, such as lytic and lysogenic growth in phage lambda.

While stochastic simulations may seem closer to molecular reality, the models discussed so far require knowledge of the reactions and the joint probability distribution of time intervals and reactions. Furthermore, the corresponding simulations are very time-consuming. Thus the models may be advantageous for fine-grained analysis but for longer time-scale and more global effects, the continuous models may be sufficient [92]. In the future, more complete models and software tools may in fact combine both formalisms into hybrid hierarchical models where fine-grained models could be used to help guide the selection of higher-lever models, and vice versa.

Probabilistic models: Bayesian networks

A different general class of probabilistic models is provided by the theory of graphical models in statistics, and in particular Bayesian or belief networks [98]. The basic idea behind graphical models is to approximate a complex multi-dimensional probability distribution using a product of simpler local probability distributions. Indeed, to a large extent, the bulk of the dependencies in many real world domains is local and therefore a global intractable high-dimensional distribution can often be approximated by a product of distributions over smaller clusters of variables defined over smaller tractable spaces [59, 99, 100]. In standard first-order Markov models of time series, for instance, the state at time t is independent of any state at times before $t-1$, conditioned on the state at time $t-1$. Thus the global distribution can be factored as a product of local distribution of the form $P(X_t|X_{t-1})$ along the corresponding chain.

More generally, a Bayesian network model is based on a directed acyclic graph (DAG) with N nodes. To each node i is associated a random variable X_i. In a genetic regulatory system, the nodes may represent genes, proteins, or other elements and the random variables X_i levels of activity, for instance in binary fashion in the most simple case. The parameters of the model are the local conditional distributions of each random variable given the random variables associated with the parent nodes

$$P(X_1,\ldots,X_N) = \prod_i P(X_i|X_j: j \in N^-(i)) \qquad (8.21)$$

$P(X_i|X_j: j \in N^-(i))$ where $N^-(i)$ denotes all the parents of vertex i. Here the parents of node i could be interpreted as the "direct" regulators of i. The Markovian independence assumption of a Bayesian network states that, conditioned on the state of its parent (immediate past), the state of a node (present) is independent of the state of all its non-descendants (distant past). This independence assumption is equivalent to the global factorization

$$P(X_1,\ldots,X_N) = \prod_{i=1}^N P(X_i|X_j: j \in N^-(i)) \qquad (8.22)$$

Given a data set D representing, for instance, expression levels derived using DNA microarray experiments, it is possible to use learning techniques [101] with heuristic approximation methods to try to infer the network architecture and parameters. Because data is still often limited and insufficient to completely determine a single model, several authors [44, 102, 103, 104, 105,

106] have developed heuristics for learning classes of models rather than single models, for focusing on particular relationships, such as causal relationships, or for leaning subnetworks, for instance from perturbed expression profiles, for a specific pathway or set of co-regulated genes.

In these approaches, Bayesian networks are used to build relatively high-level models of the data, somewhat remote from the level of molecular interactions. However, although they do not provide a direct window on the interactions, these Bayesian networks can identify causal relationships and explain certain correlations. They also come with a solid statistical foundation for handling noise, missing data, and inference. The requirement that the underlying graph be acyclic, in sharp contrast with the massively recurrent architecture of regulatory networks, has not been a major limitation so far for multiple reasons. The models are meant to capture only a subset of the dependencies at a given time. In addition, in many cases, DAG propagation methods seem to work reasonably well in networks with directed cycles [107, 108]. Most current applications of Bayesian networks to gene expression data are static in the sense that temporal dynamics is not taken into consideration. This, however, is not an intrinsic limitation of Bayesian networks *per se*. Hidden Markov models, for instance, are a class of Bayesian networks routinely used to model temporal data. Thus there exist dynamic Bayesian network models capable of incorporating temporal effects, although these would require even more data for proper model fitting. On a more speculative note, a significant interest of Bayesian networks for the future could reside in their ability to combine heterogeneous data sets, for instance microarray data with mutation data and/or epidemiological data, to produce expert systems for diagnostic and other applications.

Finally, there are other formalisms for modeling regulatory networks that are worth mentioning and that we have not covered in detail both for lack of space and because so far they seem to have had a more limited impact. These include Petri nets [109], process algebra [110], rule-based formalisms [111, 112], and grammars [113, 114, 115].

Software and databases

A condition for continued progress in systems biology will require a more thorough understanding of complex molecular interactions, such as those involved in gene transcription and regulation, and how polymeric molecules can assemble to form truly remarkable molecular processors. In addition, progress will require continued development of high-throughput methods, beyond DNA gene expression microarrays, to gather detailed measurements of complex high-dimensional biological systems. These include cDNA microarrays to determine protein–DNA interactions [116, 117], and to determine protein–protein interactions using proteomic methods such as mass spectrometry and yeast two-hybrid systems [118, 119, 120]. Additional valuable information should come from perturbation and synthetic biology, e.g., bioengineering regulatory systems *in vivo* or *in vitro*, such as the synthesis of oscillatory networks of transcriptional regulators [34, 121, 122].

Integration of all of this data and information will require proper simulation/modeling software and databases. On the modeling side, several software packages have been developed for each one of the formalisms described above. These are described in detail in the references and web sites in Appendix C. In the near future, no package is likely to satisfy the needs of all the modelers and parallel efforts with different emphases and granularity levels will have to coexist until a better understanding emerges. Because of software heterogeneity and rapid evolution, some groups are attempting to create a common overarching infrastructure for sharing data, models, results, and other resources across different projects [123].

On the database side, multiple efforts are trying to bridge the gap between genomic and metabolic/pathway data to create scalable databases to represent, store, and navigate most of the available information about the biochemical world. An object-oriented perspective is at the core of most current attempts to represent molecular information. While there are variations with each implementation, not surprisingly two fundamental objects common to almost any implementation are: (1) molecules (e.g., protein) and their attributes (e.g., phosphorylated) (2) interactions and their attributes. Sequences of adjacent entities and interactions can be assembled into pathways, which often have a graphical representation with the caveats already raised.

Being able to access pathway information should be valuable in assigning and understanding gene function and in drug discovery projects [124]. Identifying protein/genes with new functions cannot be based on the sole analysis of sequence and structure homologies, but requires taking into account information on functional groupings/pathways, gene regulation, genome structure and phylogeny, and gene location can help identify interactions and pathways. This is in essence the strategy used in [125, 126, 127] to infer a database of protein–protein interactions (see also [128, 129] as well as [130] for a brief review). Other important approaches attempt to extract information about interactions automatically from text to the existing literature [131, 132, 133]. Although the situation ought to improve in time, current databases of protein–protein interactions derived through different approaches appear to be quite noisy, incomplete, and to often disagree for multiple reasons ranging from imperfections in the methods to the complexity of the definition of the interactions themselves. Additional examples of systems biology databases, aimed at pathways or gene regulation, include aMAZE [134], EcoCyc [8], GeneNet [135], KEGG [136], TRANSFAC [137], and RegulonDB [138].

Pathway databases, together with the proper querying and graph algorithm, should allow one to progressively address new questions on a large-scale, such as comparing pathways within and across organisms, aligning pathways, completing pathways with missing information, inferring pathway phylogenies, predicting pathways in newly sequenced organisms, and ranking pathways as in finding the shortest pathway connecting two molecular species. Important similarities and differences between pathways across organisms are only beginning to be studied systematically (e.g., the lysine pathway in *Saccharomyces cerevisiae* yeast is quite different from the lysine pathway in *E. coli*). Probabilistic methods to handle noise, uncertainty, gaps in knowledge, and missing values should also play an important role in tandem with the databases.

The search for general principles

Even a complete computer simulation of the networks of a cell that can replicate and extensively predict behavior under any perturbation would not be satisfactory since, by itself, it would not provide us with much understanding of the underlying systems. What we would like to discover, of course, are the general principles of organization of these circuits. One overly pessimistic view is that either such principles do not really exist, the circuits being a morass of hacks produced by evolutionary tinkering, and/or that the complexity and stochastic variability of biological networks may preclude their complete molecular description and understanding, possibly in connection with issues of chaos and under fundamental limitations. Such a pessimistic view is premature, and in fact contradicted by current evidence. Our knowledge of gene regulation, for instance, is far from complete but this is not to say that it lacks in principles or sophistication.

To some extent, there must be some loose hierarchical organization of gene regulation. This is becoming quite clear in bacteria, with at least three broad levels of gene regulation emerging in *E. coli* [24]: (1) global control by chromosome structure (e.g., control of basal level of expression through

DNA supercoiling mechanisms correlated with [ATP]/[ADP] ratio); (2) global control of stimulons and regulons; and (3) operon-specific controls. There appears also to be a modular [139] organization of biochemical processes and gene regulation which, at least in some cases, is reflected in the modular organization of promoter regions [28]. The spatial arrangement of modules and genes along chromosomes is often tightly coupled with temporal and spatial patterns of expression. Our current difficulties are more due to the fact that this hierarchical and modular organization remains very fuzzy since we have only explored the tip of the iceberg and many more examples need to be studied in detail.

When we look at the interactions within and between modules, we can recognize principles of design well known in control theory including the extensive use of feedback control where inhibitory loops are used to provide robustness and stability, i.e., to keep parameters within a certain range, and positive loops for amplification and rapid decision between pairs of stable states. Positive feedback loops are used, for instance, to drive a cell into or away from mitosis [140], and to drive unidirectional developmental stages. Self-reinforcing loops of one kind or another[1] appear to entirely orchestrate the control of the *endo16* gene in sea urchin during development [28]. Negative feedback loops, on the other hand, play a key role in tracking circuits such as those involved in bacterial chemotaxis allowing the detection of small variations in the input signal over several orders of magnitude. The importance and extent of negative feedback is striking. For instance roughly 40% of known transcription factors in *E. coli* seem to negatively autoregulate themselves. Another well-known technique to achieve robustness is various forms of redundancy ranging from gene duplication, to multiple pathways for achieving the same results.

Indeed robustness [34, 141, 142, 143, 144, 145] is a hallmark of biological circuits. To a first approximation, regulatory networks appear to be robust, both in the sense of internal robustness with respect to biochemical parameters and concentrations and external robustness with respect to environmental variations. They are probably less robust with respect to variations in network structure. Robustness to noise comes also with the ability to exploit noise or random fluctuations [32, 33], for instance during development in order to randomize phenotypic outcomes. Robustness however is not rigidity, and comes together with the remarkable evolvability properties of biological networks. Quantitative understanding of the tradeoffs

---

[1] Two major kinds of self-reinforcing loop are seen when gene *A* activates gene *B* and vice versa, or when gene *A* activates gene *B*, and gene *B* inhibits a repressor of gene *A* (push–pull switch). Furthermore, reinforcement can occur within a cell or across different cells.

between robustness and evolvability remains one of the theoretical challenges. Comparative genomics and *in vitro* evolution may shed light on the evolution of biochemical circuits, and provide a better understanding of the constraints they must satisfy and the robustness/evolvability tradeoff.

It is also clear that we need to develop a better understanding of DNA structure [146, 147]. DNA structural properties have been increasingly implicated in regulatory phenomena and this is not too surprising given that the cellular regulatory machinery must act on thousands of genes sparsely hidden in the tangle of chromatin fibers. While DNA structure is likely to play a role at multiple levels and scales, most of the information available today is at the local level, from the structural properties of particular binding sites, such as the bendability of TATA box region [148, 149, 150], to the fact that transcription co-activators such as CBP have histone acetyltransferase activity and are likely to modulate chromatin structure during transcription in eukaryotes [27, 151, 152, 153, 154]. Other mechanisms involving DNA supercoiling have been implicated in transcription regulation [155]. Unfortunately, our understanding of DNA structure seems to lag our already incomplete understanding of protein structure, even on short length scales, partly due to the peculiarities of the DNA double helix, in particular its considerable length coupled with small transversal size.

As already discussed, biochemical networks often have a graphical representation. An interesting approach to deriving large-scale principles for these networks is to look for scaling laws [156] in the corresponding graphs and study how they deviate from completely random graphs [157, 158]. In traditional random graphs [159], where each edge is present or absent with a fixed probability, the probability for a node to be associated with k edges falls off exponentially with k. In many graphs associated with phenomena as diverse as the World Wide Web or social organization, the connectivity appears to be random but follows a different power-law falling off with a power law $k^{-\gamma}$ rather than an exponential. These scale-free networks are also characterized by the presence of a few highly connected nodes (hubs) and the so-called small-world effect according to which there is always a relatively short path connecting any two nodes in the graph. In general, the diameter of the graph, i.e., the average length of the shortest path between any two nodes, grows logarithmically with the number of the nodes, provided the average connectivity of a node is fixed. Analysis of connectivity in metabolic networks of several dozen organisms inferred from the WIT database in [158], seem to show that metabolic networks in all three domains of life have power scaling laws. For instance, in the case of *E. coli* the exponent γ for both incoming and outgoing edges is close to 2.2. The diameter, however, appear to be constant across organisms, irrespective of the number of substrates. The latter suggests that the average connectivity of substrates increases with organism complexity. A constant diameter may be an important survival and growth advantage since a larger diameter could diminish the organism's ability for efficient response to external changes or internal errors. Removal of a random node in general does not affect the diameter, another sign of robustness. The metabolic hubs appear also to be conserved across organisms. This is analogous to the equivalent observation in protein networks that the proteins that interact with many other proteins, such as histones, actin and tubulin, have evolved very little. It is reasonable to suspect that similar results hold for other biochemical networks, and for regulatory networks in particular. Ranking of connectivity of transcription factors in *E. coli* ought to exhibit power-law behavior. While not all the necessary data may yet be available, DNA microarrays ought to help rapidly fill the remaining gaps in the graph of regulatory influences.

Thus, although we are in search for new paradigms, the situation may be far from hopeless. In fact, the problem may be primarily one of size. We need many more good detailed models of modules, and entire databases of modules – both lengthy and painstaking tasks – in order to pierce through the evolutionary tinkering fog. When learning music, it is always better to master one key rather than trying to learn all keys simultaneously. Likewise, for the near future, most progress is still likely to come from the detailed study of specific circuits or subnetworks, bringing to bear all available technologies, from conventional studies to high-throughput to modeling, on single well-defined system. DNA microarrays should continue to play a key role in this endeavor.

REFERENCES

1. Noble, D. Modeling the heart: From genes to cells to the whole organ. 2002. *Science* 295:1669–1682.
2. Kitano, H. Systems biology: A brief overview. 2002. *Science* 295:1662–1664.
3. Ideker, T., Thorsson, V., Ranish, J. A., Christmas, R., Buhler, J., Eng, J. K., Bumgarner, R., Goodlett, D. R., Aebersold, R., and Hood, L. Integrated genomic and proteomic analysis of a systematically perturbed metabolic network. 2001. *Science* 292:929–934.
4. Loomis, W. F., and Sternberg, P. W. Genetic networks. 1995. *Science* 269:649.
5. Thieffry, D. From global expression data to gene networks. 1999. *BioEssays* 21(11):895–899.
6. Salgado, H., Santos-Zavaleta, A., Gama-Castro, S., Millán-Zárate, D.,

Blattner, F. R., and Collado-Vides, J. RegulonDB (version 3.0): transcriptional regulation and operon organization in *Escherichia coli* K-12. 2000. *Nucleic Acids Research* 28(1):65–67.

7. Edwards, J. S., and Palsson, B. O. The *Escherichia coli* MG1655 *in silico* metabolic genotype: its definition, characteristics, and capabilities. 2000. *Proceedings of the National Academy of Sciences of the USA* 97(10):5528–5533.

8. Karp, P. D., Riley, M., Paley, S. M., Pellegrini-Toole, A., and Krummenacker, M. EcoCyc: encyclopedia of *Escherichia coli* genes and metabolism. 1999. *Nucleic Acids Research* 27(1):55–58.

9. Thieffrey, D., Huerta, A. M., Pérez-Rueda, E., and Colladio-Vides, J. From specific gene regulation to genomic networks: a global analysis of transcriptional regulation in *Escherichia coli*. 1998. *BioEssays* 20:433–440.

10. Endy, D., and Brent, R. Modelling cellular behavior. 2001. *Nature* 409:391–395.

11. Hasty, J., McMillen, D., Isaacs, F., and Collins, J. J. Computational studies of gene regulatory networks: *in numero* molecular biology. 2001. *Nature Reviews of Genetics* 2:268–279.

12. Smolen, P., Baxter, D. A., and Byrne, J. H. Frequency selectivity, multistability, and oscillations emerge from models of genetic regulatory systems. 1998. *American Journal of Physiology* 274:C531–C542.

13. Bower, J. M., and Bolouri, H. (eds.) *Computational Modeling of Genetic and Biochemical Networks*. 2001. MIT Press, Cambridge, MA.

14. Schlick, T., Skeel, R., Brünger, A., Kalé, L., Board, J. A. Jr., Hermans, J., and Schulten, K. Algorithmic challenges in computational molecular biophysics. 1999. *Journal of Computational Physics* 151:9–48.

15. Umbarger, H. E. Evidence for a negative feedback mechanism in the biosynthesis of isoleucine. 1956. *Science* 123:848.

16. Monod, J., Wyman, J., and Changeux, J. P. On nature of allosteric transitions: a plausible model. 1965. *Journal of Molecular Biology* 12:88.

17. Koshland, D. E., Nemethy, G., and Filmer, D. Comparison of experimental binding data and theoretical models in proteins containing subunits. 1966. *Biochemistry* 5:365.

18. Goodsell, D. S. *The Machinery of Life*, reprinted edn. 1998. Copernicus Books, New York.

19. Levchenko, A., Bruck, J., and Sternberg, P. W. Scaffold proteins may biphasically affect the levels of mitogen-activated protein kinase signaling and reduce its threshold properties. 2000. *Proceedings of the National Academy of Sciences of the USA* 97:5818–5823.

20. Edwards, J. S., and Palsson, B. O. The *Escherichia coli* MG1655 *in silico* metabolic genotype: Its definition, characteristics, and capabilities. 2000. *Proceedings of the National Academy of Sciences of the USA* 97:5528–5533.

21. Bray. D. Protein molecules as computational elements in living cells. 1995. *Nature* 376:307–312.

22. Claverie, J. M. Gene number: what if there are only 30000 human genes? 2001. *Science* 291:1255–1257.

23. Maniatis, T., and Reed, R. An extensive network of coupling among gene expression machines. 2002. *Nature* 416:499–506.

24. Shaechter, M., and The View From Here Group. *Escherichia coli* and *salmonella* 2000: the view from here. 2001. *Microbiology and Molecular Biology Reviews* 65:119–130.

25. Blackman, R. K., Sanicola, M., Raftery, L. A., Gillevet, T., and Gelbart,

W. M. An extensive 3′ *cis*-regulatory region directs the imaginal disk expression of decapentaplegic, a member of the TGF-beta family in *Drosophila*. 1991. *Development* 111;657–666.

26. van den Heuvel, M., Harryman-Samos, C., Klingensmith, J., Perrimon, N., and Nusse, R. Mutations in the segment polarity genes wingless and porcupine impair secretion of the wingless protein. 1993. *EMBO Journal* 12:5293–5302.

27. Latchman, D. S. *Eukaryotic Transcription Factors*, 3rd edn. 1998. Academic Press, San Diego, CA.

28. Yuh, C. H., Bolouri, H., and Davidson, E. H. *Cis*-regulatory logic in the *endo 16* gene: switching from a specification to a differentiation mode of control. 2001. *Development* 128:617–629.

29. McAdams, H. H., and Shapiro, L. Circuit simulation of genetic networks. 1995. *Science* 269:650–656.

30. Yuh, C. H., Bolouri, H., and Davidson, E. H. Genomic *cis*-regulatory logic: experimental and computational analysis of a sea urchin gene. 1998. *Science* 279:1896–1902.

31. Davidson, E. H. *et al*. A genomic regulatory network for development. 2002. *Science* 295:1669–1678.

32. McAdams, H. H., and Arkin, A. It's a noisy business! genetic regulation at the nanomolar scale. 1999. *Trends in Genetics* 15:65–69.

33. Hasty, J., Pradines, J., Dolnik, M., and Collins, J. J. Noise-based switches and amplifiers for gene expression. 2000. *Proceedings of the National Academy of Sciences of the USA* 97:2075–2080.

34. Becksel, A., and Serrano, L. Engineering stability in gene networks by autoregulation. 2000. *Nature* 405:590–593.

35. Barkal, N., and Liebler, S. Robustness in simple biochemical networks. *Nature* 387:913–917.

36. Barkai, N., and Liebler, S. Biological rhythms: circadian clocks limited by noise. 2000. *Nature* 403:267–268.

37. Kauffman, S. A. Metabolic stability and epigenesis in randomly constructed genetic nets. 1969. *Journal of Theoretical Biology* 22:437–467.

38. Kauffman, S. A. The large-scale structure and dynamics of gene control circuits: an ensemble approach. 1974. *Journal of Theoretical Biology* 44:167–190.

39. Kauffman, S. A. Requirements for evolvability in complex systems: orderly dynamics and frozen components. 1990. *Physica D* 42:135–152.

40. Mjolsness, E., Sharp, D. H., and Reinitz, J. A connectionist model of development. 1991. *Journal of Theoretical Biology* 152:429–453.

41. Voit, E. O. *Canonical Nonlinear Modeling*. 1991. Van Nostrand and Reinhold, New York.

42. Savageau, M. A. Power-law formalism: a canonical nonlinear approach to modeling and analysis. In V. Lakshmikantham, editor, *World Congress of Nonlinear Analysts 92*, vol. 4, pp. 3323–3334. 1996. Walter de Gruyter Publishers, Berlin.

43. Hlavacek, W. S., and Savageau, M. S. Completely uncoupled and perfectly coupled gene expression in repressible systems. 1997. *Journal of Molecular Biology* 266:538–558.

44. Friedman, N., Linial, M., Nachman, I., and Pe'er, D. Using Bayesian networks to analyze expression data. 2000. *Journal of Computational Biology* 7:601–620.

45. McAdams, H. H., and Arkin, A. Simulation of prokaryotic genetic circuits. 1998. *Annual Reviews of Biophysical and Biomolecular Structures* 27:199–224.

46. Smolen, P., Baxter, D. A., and Byrne, J. H. Modeling transcriptional control in gene networks: methods, recent results, and future directions. 2000. *B. Math. Biol.*, 62:247–292.

47. Kauffman, S. A. Homeostasis and differentiation in random genetic control networks. 1969. *Nature* 224:177–178.

48. Kauffman, S. A. Gene regulation networks: a theory for their global structure and behaviors. In A. N. Other, editor, *Current Topics in Developmental Biology*, vol. 6, pp. 145–182. 1977. Academic Press, New York.

49. Thomas, R. Logical analysis of systems comprising feedback loops. 1978. *Journal of Theoretical Biology* 73:631–656.

50. Thomas, R. Boolean formalization of genetic control circuits. 1973. *Journal of Theoretical Biology* 42:563–585.

51. Thomas, R., d'Ari, R. *Biological Feedback*. 1990. CRC Press, Boca Raton, FL.

52. Thomas, R. Regulatory networks seen as asynchronous automata: a logical description. 1991. *Journal of Theoretical Biology* 153:1–23.

53. Baldi, P., and Atiya, A. F. Oscillations and synchronizations in neural networks: an exploration of the labeling hypothesis. 1989. *International Journal of Neural Systems* 1(2):103–124.

54. Baldi, P., and Atiya, A. F. How delays affect neural dynamics and learning. 1994. *IEEE Transactions on Neural Networks* 5(4):626–635.

55. Arik, S. Stability analysis of delayed neural networks. 2000. *IEEE Transactions on Circuits and Systems* I, 47:1089–1092.

56. Joy, M. P. Results concerning the absolute stability of delayed neural networks. 2000. *Neural Networks* 13:613–616.

57. Hornik, K., Stinchcombe, M., and White, H. Universal approximation of an unknown function and its derivatives using multilayer feedforward networks. 1990. *Neural Networks* 3:551–560.

58. Hornik, K., Stinchcombe, M., White, H., and Auer, P. Degree of approximation results for feedforward networks approximating unknown mappings and their derivatives. 1994. *Neural Computation* 6:1262–1275.

59. Baldi, P., and Brunak, S. *Bioinformatics: The Machine Learning Approach*. 2001. MIT Press, Cambridge, MA.

60. Savageau, M. A. Enzyme kinetics *in vitro* and *in vivo*: Michaelis–Menten revisited. In E. E. Bittar, editor, *Principles of Medical Biology*, vol. 4, pp. 93–146. 1995. JAI Press Inc., Greenwich, CT.

61. Baldi, P. Gradient descent learning algorithms overview: a general dynamical systems perspective. 1995. *IEEE Transactions on Neural Networks* 6(1):182–195.

62. Fraser, C. M., Gocayne, J. D., White, O., Adams, M. D., Clayton, R. A., Fleischmann, R. D., Bult, C. J., Kerlavage, A. R., Sutton, G., Kelley, J. M., *et al*. The minimal gene complement of *Mycoplasma genitalium*. 1995. *Science* 270:397–403.

63. Snoussi. E. H. Qualitative dynamics of piecewise-linear differential equations: a discrete mapping approach. 1989. *Dynamics and Stability of Systems* 4(3–4):189–207.

64. Plahte, E., Mestl, T., and Omholt, S. W. A methodological basis for description and analysis of systems with complex switch-like interactions. 1998. *Journal of Mathematical Biology* 36:321–348.

65. Edwards, R., and Glass, L. Combinatorial explosion in model gene networks. 2000. *Chaos* 10(3):691–704.

66. Edwards, R., Seigelmann, H. T., Aziza, K., and Glass, L. Symbolic dynamics and computation in model gene networks. 2001. *Chaos* 11(1):160–169.
67. Lewis, J. E., and Glass, L. Steady states, limit cycles, and chaos in models of complex biological networks. 1991. *International Journal of Bifurcation and Chaos* 1(2):477–483.
68. Mestl, T., Plahte, E., and Omholt, S. W. Periodic solutions in systems of piecewise-linear differential equations. 1995. *Dynamics and Stability of Systems* 10(2):179–193.
69. Plahte, E., Mestl, T., and Omholt, S. W. Stationary states in food web models with threshold relationships. 1995. *Journal of Biological Systems* 3(2):569–577.
70. Mestl, T., Plahte, E., and Omholt, S. W. A mathematical framework for describing and analysing gene regulatory networks. 1995. *Journal of Theoretical Biology* 176:291–300.
71. Snoussi, E. H., and Thomas, R. Logical identification of all steady states: the concept of feedback loop characteristic states. 1993. *B. Math. Biol.*, 55(5):973–991.
72. Prokudina, E. I., Valeev, R. Y., and Tchuraev, R. N. A new method for the analysis of the dynamics of the molecular genetic control systems. II. Application of the method of generalized threshold models in the investigation of concrete genetic systems. 1991. *Journal of Theoretical Biology* 151:89–110.
73. Omholt, S. W., Kefang, X., Anderson, Ø., and Plahte, E. Description and analysis of switchlike regulatory networks exemplified by a model of cellular iron homeostasis. 1998. *Journal of Theoretical Biology* 195:339–350.
74. Sanchez, L., van Helden, J., and Thieffrey, D. Establishment of the Dorso-ventral pattern during embryonic development of *Drosophila melanogaster*: a logical analysis. 1997. *Journal of Theoretical Biology* 189:377–389.
75. de Jong, H., Page, M., Hernandez, C., and Geiselmann, J. Qualitative simulation of genetic regulatory networks: method and application. In B. Nebel, editor, *Proceedings of the 17th International Joint Conference on Artificial Intelligence, IJCAI-01*, pp. 67–73. 2001. Morgan Kauffman, San Mateo, CA.
76. Britton, N. F. *Reaction–Diffusion Equations and Their Applications to Biology*. 1986. Academic Press, London.
77. Lacalli, T. C. Modeling the *Drosophila* pair-rule pattern by reaction diffusion: gap input and pattern control in a 4-morphogen system. 1990. *Journal of Theoretical Biology* 144:171–194.
78. Kauffman, S. A. *The Origins of Order: Self-Organization and Selection in Evolution*. 1993. Oxford University Press, New York.
79. Maini, P. K., Painter, K. J., and Nguyen Phong Chau, H. Spatial pattern formation in chemical and biological systems. 1997. *Journal of the Chemical Society, Faraday Transactions* 93(20):3601–3610.
80. Turing, A. M. The chemical basis of morphogenesis. 1951. *Philosophical Transactions of the Royal Society B* 237:37–72.
81. Nicolis, G., and Prigogine, I. *Self-Organization in Nonequilibrium Systems: From Dissipative Structures to Order Through Fluctuations*. 1977. Wiley-Interscience, New York.
82. Segel, L. A. *Modeling Dynamic Phenomena in Molecular and Cellular Biology*. 1984. Cambridge University Press, Cambridge, UK.
83. Meinhardt, H. Pattern formation by local self-activation and lateral inhibition. 2000. *BioEssays* 22(8):753–760.

84. Lacalli, T. C., Wilkinson, D. A., and Harrison, L. G. Theoretical aspects of stripe formation in relation to *Drosophila* segmentation. 1988. *Development* 103:105–113.

85. Hunding, A., Kauffman, S. A., and Goodwin, B. C., *Drosophila* segmentation: supercomputer simulation of prepattern hierarchy. 1990. *Journal of Theoretical Biology* 145:369–384.

86. Marnellos, G., Deblandre, G. A., Mjolsness, E., and Kintner, G. Delta-Notch lateral inhibitory patterning in the emergence of ciliated cells in *Xenopus*: experimental observations and a gene network model. In R. B. Altman, K. Lauderdale, A. K. Dunker, L. Hunter, and T. E. Klein, editors, *Proceedings of the Pacific Symposium on Biocomputing (PSB 2000)*, vol. 5, pp. 326–337. 2000. World Scientific Publishing, Singapore.

87. Reinitz, J., Mjolsness, E., and Sharp, D. H. Model for cooperative control of positional information in *Drosophila* by bicoid and maternal hunchback. 1995. *Journal of Experimental Zoology* 271:47–56.

88. Sharp, D. H., and Reinitz, J. Prediction of mutant expression patterns using gene circuits. 1998. *BioSystems* 47:79–90.

89. Reinitz, J. and Sharp, D. H. Gene circuits and their uses. In J. Collado-Vides, B. Magasanik, and T. F. Smith, editors, *Integrative Approaches to Molecular Biology*, pp. 253–272. 1996. MIT Press, Cambridge, MA.

90. Myasnikova, E., Samsonova, A., Kozlov, K., Samsonova, and Reinitz, J. Registration of the expression patterns of *Drosophila* segmentation genes by two independent methods. 2001. *Bioinformatics* 17(1):3–12.

91. van Kampen, N. G. *Stochastic Processes in Physics and Chemistry*, revised edn. 1997. Elsevier, Amsterdam.

92. Gillespie, D. T. The chemical Langevin equation. 2000. *Journal of Chemical Physics* 113(1):297–306.

93. Gillespie, D. T. Exact stochastic simulation of coupled chemical reactions. 1977. *Journal of Physical Chemistry* 81(25):2340–2361.

94. Gibson, M. A., and Bruck, J. Efficient exact stochastic simulation of chemical systems with many species and many channels. 2000. *Journal of Physical Chemistry A* 104:1876–1889.

95. McAdams, H. H., and Arkin, A. Stochastic mechanism in gene expression. 1997. *Proceedings of National Academy of Sciences of the USA* 94:814–819.

96. Arkin, A., Ross, J., and McAdams, H. A. Stochastic kinetic analysis of developmental pathway bifurcation in phage λ-infected *Escherichia coli* cells. 1998. *Genetics* 149:1633–1648.

97. Barkai, N., and Liebler, S. Circadian clocks limited by noise. 1999. *Nature* 403:267–268.

98. Pearl, J. *Probabilistic Reasoning in Intelligent Systems: Networks of Plausible Inference*. 1988. Morgan Kaufmann, San Mateo, CA.

99. Lauritzen, S. L. *Graphical Models*. 1996. Oxford University Press, Oxford, UK.

100. Jordan, M. I. (ed.). *Learning in Graphical Models*. 1999. MIT Press, Cambridge, MA.

101. Heckerman, D. A tutorial on learning with Bayesian networks. In M. I. Jordan, editor, *Learning in Graphical Models*, pp. 00–00. 1998. Kluwer, Dordrecht.

102. van Someren, E. P., Wessels, L. F. A., and Reinders, M. J. T. Linear modeling of genetic networks from experimental data. In *Proceedings of the 2000 Conference on Intelligent Systems for Molecular Biology (ISMB00)*, La Jolla, CA, pp. 355–366. 2000. AAAI Press, Menlo Park, CA.

103. Zien, A., Kuffner, R., Zimmer, R., and Lengauer, T. Analysis of gene

expression data with pathway scores. In *Proceedings of the 2000 Conference on Intelligent Systems for Molecular Biology (ISMB00), La Jolla, CA*, pp. 407–417. 2000. AAAI Press, Menlo Park, CA.

104. Elidan, G., Peer, D., Regev, A., and Friedman, N. Inferring subnetworks from perturbed expression profiles. 2001. *Bioinformatics* 17:S215–S224. Supplement 1.

105. Gasch, A., Friedman, N., Segal, E., Taskar, B., and Koller, D. Rich probabilistic models for gene expression. 2001. *Bioinformatics* 17:S243–S252, Supplement 1.

106. Tanay, A., and Shamir, R. Computational expansion of genetic networks. 2001. *Bioinformatics* 17:S270–S278, Supplement 1.

107. Weiss, Y. Correctness of local probability propagation in graphical models with loops. 2000. *Neural Computation* 12:1–41.

108. Yedidia, J. S., Freeman, W. T., and Weiss, Y. Generalized belief propagation. In T. Leen, T. Dietterich, and V. Tresp, editors, *Advances in Neural Information Processing Systems*, vol. 13 (Proceedings of the NIPS 2000 Conference). 2001. MIT Press, Boston, MA.

109. Goss, P. J. E., and Peccoud, J. Quantitative modeling of stochastic systems in molecular biology by using stochastic Petri nets. 1998. *Proceedings of the National Academy of Sciences of the USA* 95:6750–6755.

110. Regev, A., Silverman, W., and Shapiro, E. Representation and simulation of biochemical processes using the π-calculus process algebra. In R. B. Altman, A. K. Dunker, L. Hunter, K. Lauderdale, and T. E. Klein, editors, *Proceedings of the Pacific Symposium on Biocomputing* (PSB 2001), vol. 6, pp. 459–470.

111. Meyers, S., and Friedland, P. Knowledge-based simulation of genetic regulation in bacteriophage lambda. 1984. *Nucleic Acids Research* 12(1):1–9.

112. Shimada, T., Hagiya, M., Arita, M., Nishizaki, S., and Tan, C. L. Knowledge-based simulation of regulatory action in lambda phage. 1995. *International Journal of Artificial Intelligence Tools* 4(4):511–524.

113. Lindenmayer, A. Mathematical models for cellular interaction in development. I. Filaments with one-sided inputs. 1968. *Journal of Theoretical Biology* 18:280–289.

114. Collado-Vides, J., Guitièrrez-Ríos, R. M., and Bel-Enguix, G. Networks of transcriptional regulation encoded in a grammatical model. 1998. *BioSystems* 47:103–118.

115. Hofestädt, R., and Meineke, F. Interactive and modelling and simulation of biochemical networks. 1995. *Computers in Biology and Medicine* 25(3):321–334.

116. Ren, B., Robert, F., Wyrick, J. J., Aparicio, O., Jennings, E. G., Simon, I., Zeitlinger, J., Schreiber, J., Hannett, N., Kanin, E., Volkert, T. L., Wilson, C. J., Bell, S. P., and Young, R. A. Genome-wide location and function of DNA binding proteins. 2000. *Science* 290:2306–2309.

117. Iyer, V. R., Horak, C. E., Scafe, C. S., Botstein, D., Snyder, M., and Brown, P. O. Genomic binding sites of the yeast cell-cycle transcription factors SBF and MBF. 2001. *Nature* 409:533–538.

118. Pandey, A., and Mann, M. Proteomics to study genes and genomes. 2000. *Nature* 405:837–846.

119. Zhu, H., and Snyder, M. Protein arrays and microarrays. 2001. *Current Opinion in Chemical Biology* 5:40–45.

120. Tucker, C. L., Gera, J. F., and Uetz, P. Towards an understanding of complex protein networks. 2001. *Trends in Cell Biology* 11(3):102–106.

121. Gardner, T. S., Cantor, C. R., and Collins, J. J. Construction of a genetic toggle switch in *Escherichia coli*. 2000. *Nature* 403:339–342.
122. Elowitz, M., and Liebler, S. A synthetic oscillatory network of transcriptional regulators. 2000. *Nature* 403:335–338.
123. Systems Biology Workbench Project. http://www.cds.caltech.edu/erato/
124. Karp, P. D., Krummenacker, M., Paley, S., and Wagg, J. Integrated pathway–genome databases and their role in drug discovery. 1999. *Trends in Biotechnology* 17:275–281.
125. Marcotte, E. M., Pellegrini, M., Ng, H. L., Rice, D. W., Yeates, T. O., and Eisenberg, D. Detecting protein function and protein–protein interactions from genome sequences. 1999. *Science* 285:751–753.
126. Xenarios, I., Fernandez, E., Salwinsky, L., Duan, X. J., Thompson, M. J., Marcotte, E. M., and Eisenberg, D. DIP: the database of interacting proteins: 2001 update. 2001. *Nucleic Acids Research* 29:239–241.
127. Wojcik, J., and Schachter, V. Protein–protein interaction map inference using interacting domain profile pairs. 2001. *Bioinformatics* 17:S296–S305, Supplement 1.
128. Pazos, F., and Valencia, A. Similarity of phylogenetic trees as indicator of protein–protein interaction. 2001. *Protein Engineering* 14:609–614.
129. Niggemann, O., Lappe, M., Park, J., and Holm, L. Generating protein interaction maps from incomplete data: application to fold assignment. 2001. *Bioinformatics* 17:S149–S156, Supplement 1.
130. Xenarios, I., and Eisenberg, D. Protein interaction databases. 2001. *Current Opinion in Biotechnology* 12:334–339.
131. Blaschke, C., Andrade, M. A., Ouzounis, C., and Valencia, A. Automatic extraction of biological information from scientific text: protein–protein interactions. In *Proceedings of the 1999 Conference on Intelligent Systems for Molecular Biology (ISMB99), Heidelberg, Germany*, pp. 60–67. 1999. AAAI Press, Menlo Park, CA.
132. Marcotte, E. M., Xenarios, I., and Eisenberg, D. Mining literature for protein–protein interactions. 2001. *Bioinformatics* 17:359–363.
133. Yu, H., Krauthammer, M., Friedman, C., Kra, P., and Rzhetsky, A. GENIES: a natural-language processing system for the extraction of molecular pathways from journal articles. 2001. *Bioinformatics* 17:S74–S82, Supplement 1.
134. van Helden, J., Naim, A., Mancuso, R., Eldridge, M., Wernisch, L., Gilbert, D., and Wodak, S. J. Representing and analysing molecular and cellular function using the computer. 2000. *Biological Chemistry* 381:921–935.
135. Kolpakov, F. A., Ananko, E. A., Kolesov, G. B., and Kolchanov, N. A. GeneNet: a gene network database and its automated visualization. 1999. *Bioinformatics* 14(6):529–537.
136. Kanehisa, M., and Goto, S. KEGG: Kyoto Encyclopedia of Genes and Genomes. 2000. *Nucleic Acids Research* 28(1):27–30.
137. Wingender, E., Chen, X., Fricke, E., Geffers, R., Hehl, R., Liebich, I., Krull, M., Matys, V., Michael, H., Ohnhauser, R., Pruss, M., Schacherer, F., Thiele, S., and Urbach, S. The TRANSFAC system on gene expression regulation. 2001. *Nucleic Acids Research* 29:281–284.
138. Salgado, H., Santos, A., Garza-Ramos, U., van Helden, J., Díaz, E., and Collado-Vides, J. RegulonDB (version 2.0): a database on transcriptional regulation in *Escherichia coli*. 2000. *Nucleic Acids Research* 27(1):59–60.
139. Hartwell, L. H., Hopfield, J. J., Leibler, S., and Murray, A. W. From molecular to modular cell biology. 1999. *Nature* 402 (Supp.):C47–C52.

140. Morgan, D. O. Cyclin-dependent kinases: engines, clocks, and microprocessors, 1997. *Annual Review of Cellular and Developmental Biology* 13:261–291.
141. Barkai, N., and Leibler, S. Robustness in simple biochemical networks. 1997. *Nature* 387:913–917.
142. Yi, T., Huang, Y., Simon, M. I., and Doyle, J. Robust perfect adaptation in bacterial chemotaxis through integral feedback control. 2000. *Proceedings of the National Academy of Sciences of the USA* 97:4649–4653.
143. Morohashi, M., Winn, A. E., Borisuk, M. T., Bolouri, H., Doyle, J., and Kitano, H. Robustness as a measure of plausibility. 2002. *Journal of Theoretical Biology*. (in press)
144. Csete, M. E., and Doyle, J. C. Reverse engineering of biological complexity. 2002. *Science* 295:1664–1669.
145. Houchmandzadeh, B., Wieschaus, E., and Leibler, S. Establishment of development precision and proportions in the early *Drosophila* embryo. 2002. *Nature* 415:798–802.
146. Sinden, R. R. *DNA Structure and Function*. 1994. Academic Press, San Diego, CA.
147. Baldi, P., and Baisnée, P. F. Sequence analysis by additive scales: DNA structure for sequences and repeats of all lengths. 2000. *Bioinformatics* 16(10):865–889.
148. Parvin, J. D., McCormick, R. J., Sharp, P. A., and Fisher, D. E. Pre-bending of a promoter sequence enhances affinity for the TATA-binding factor. 1995. *Nature* 373:724–727.
149. Starr, D. B., Hoopes, B. C., and Hawley, D. K. DNA bending is an important component of site-specific recognition by the TATA binding protein. 1995. *Journal of Molecular Biology* 250:434–446.
150. Grove, A., Galeone, A., Mayol, L., and Geiduschek, E. P. Localized DNA flexibility contributes to target site selection by DNA-bending proteins. 1996. *Journal of Molecular Biology* 260:120–125.
151. Pazin, M. J., and Kadonaga, J. T. SWI2/SNF2 and related proteins: ATP-driven motors that disrupt protein–DNA interactions? 1997. *Cell* 88:737–740.
152. Tsukiyama, T., and Wu, C. Chromatin remodeling and transcription. 1997. *Current Opinion in Genetics and Development* 7:182–191.
153. Werner, M. H., and Burley, S. K. Architectural transcription factors: proteins that remodel DNA. 1997. *Cell* 88:733–736.
154. Gorm Pedersen, A., Baldi, P., Brunak, S., and Chauvin, Y. DNA structure in human RNA polymerase II promoters. 1998. *Journal of Molecular Biology* 281:663–673.
155. Sheridan, S. D., Benham, C. J., and Hatfield, G. W. Activation of gene expression by a novel DNA structural transmission mechanism that requires supercoiling-induced DNA duplex destabilization in an upstream activating sequence. 1998. *Journal of Biological Chemistry* 273:21298–21308.
156. Banavar, J. R., Maritan, A., and Rinaldo, A. Size and form in efficient transportation networks. 1999. *Nature* 399:130–132.
157. Barabasi, A., and Albert, R. Emergence of scaling in random networks. 1999. *Science* 286:509–512.
158. Jeong, H., Tomber, B., Albert, R., Oltvai, Z. N., Barabasi, A.-L. The large-scale organization of metabolic networks. 2000. *Nature* 407:651–654.
159. Bollobas, B. *Random Graphs*. 1985. Academic Press, New York.

Appendix A
Experimental protocols

Total RNA isolation from bacteria

Total RNA is isolated from cells at an A_{600} of 0.5–0.6. Ten-ml samples of cultures of growing cells are pipetted directly into 10 ml of boiling lysis buffer (1% SDS, 0.1 M NaCl, 8 mM EDTA) and mixed at 100 °C for 1.5 min. These samples are transferred to 250-ml Erlenmeyer flasks, mixed with an equal volume of acid phenol (pH 4.3), and shaken vigorously for 6 min at 64 °C. After centrifugation, the aqueous phase is transferred to a fresh Erlenmeyer flask, and the hot acid phenol extraction procedure is repeated. The second aqueous phase is extracted with phenol:chloroform:isoamyl alcohol (25:24:1, pH 4.3) at room temperature and, finally, twice with chloroform-isoamyl alcohol (24:1). Total RNA is precipitated with two volumes of ethanol in 0.3 M sodium acetate (pH 5.3), washed with 70% ethanol, and redissolved in RNAase-free water. The contaminating genomic DNA is removed from the total RNA with Ambion's DNA-free kit ™ (catalog no. 1906). Since genomic DNA is a common source of high background on DNA arrays, this step is repeated at least once. The quality and integrity of the total RNA preparation is ascertained by electrophoresis in a 1.2 % agarose gel run in 1 × TAE (40 mM Tris-acetate, 2 mM EDTA) buffer. Appropriate modifications of these methods can be used for total RNA extraction from eukaryotic cells.

cDNA synthesis for the preparation of ^{33}P-labeled bacterial or eukaryotic targets for hybridization to pre-synthesized nylon filter arrays

cDNA synthesis for the preparation of ^{33}P-labeled targets for hybridization to the nylon filters is performed with 20 μg of total RNA and 37.5 ng of random hexamer primers. To anneal the primers to the RNA, this mixture is heated at 70 °C for 3 min and quick cooled on ice in the presence of annealing buffer (1 × RT buffer: 50 mM Tris-HCl, 8 mM $MgCl_2$, 30 mM KCl, 1 mM DTT, pH 8.5). cDNA synthesis is performed at 42 °C for 3 hrs in a 60-μl reaction mixture containing the RNA and primer mixture, reverse transcriptase buffer (50 mM Tris-HCl, 8 mM $MgCl_2$, 30 mM KCl, 1 mM DTT, pH 8.5) containing: 1 mM each dATP, dGTP, and dTTP, 50 μCi [α^{33}P]-dCTP (3000 Ci/mmol), 20 units of ribonuclease inhibitor III, and 4 μl (88 units) of AMV reverse transcriptase. Labeled cDNA is separated from unincorporated nucleotides on Sephadex G25 spin columns (Roche Biochemical) in a tabletop clinical centrifuge. Three of these preparations are combined for hybridization to the DNA microarray filters.

Hybridization of ^{33}P-labeled targets to pre-synthesized nylon filter arrays

The nylon filters are soaked in $2\times$ SSPE ($20\times$ SSPE contains 3 M NaCl, 0.2 M NaH$_2$PO$_4$, 25 mM EDTA) for 10 min and prehybridized in 10 ml of prehybridization solution ($5\times$ SSPE, 2% SDS, $1\times$ Denhardt's solution ($50\times$ Denhardt's solution contains 5 g of Ficoll, 5 g of polyvinylpyrrolidone, 5 g of bovine serum albumin and H$_2$O to 500 ml) and 0.1 mg/ml of sheared herring sperm DNA) for 1 hr at 65 °C. $3-5\times 10^7$ cpm of cDNA targets in 500 µl of prehybridization solution is heated at 95 °C for 10 min, rapidly cooled on ice, and added to an additional 5.5 ml of prehybridization solution. The prehybridization solution is removed and replaced with the prehybridization solution containing the ^{33}P-labeled cDNA targets. Hybridization is carried out for 15–18 hrs at 65 °C. Following hybridization, each filter is rinsed with 50 ml of $0.5\times$ SSPE containing 0.2% SDS at room temperature for 3 min, followed by three more washes in $0.5\times$ SSPE containing 0.2% SDS solution at 65 °C for 20 min each. The filters are partially air-dried, wrapped in Saran Wrap,® and exposed to a phosphorimager screen for 15–30 hrs. Following phosphorimaging, the targets are stripped from the filters by microwaving at 30% of maximal power (1400 W) in 500 ml of 10 mM Tris solution (pH 8.0) containing 1 mM EDTA and 1% SDS for 20 min. Stripped filters are wrapped in Saran Wrap® and stored in the presence of damp paper towels in sealed plastic bags at 4 °C.

Cy3, Cy5 target labeling of RNA for hybridization to pre-synthesized glass slide arrays

The procedure for preparing fluorophor-labeled cDNA targets consists of four steps: a reverse transcription reaction, RNA hydrolysis, removal of Tris-HCl, and cDNA labeling with Cy3 or Cy5. For the reverse transcription reaction, 1 µl of a random hexamers mix (5 µg/µl) and 20 µg of total RNA (up to 14.5 µl) of control or experimental samples is mixed in a PCR tube and heated to 70 °C for 10 min and rapidly cooled to 4 °C. For cDNA synthesis, 3 µl of $10\times$ RT buffer (500 mM Tris-HCl (pH 8.3), 750 mM KCl, 30 mM MgCl$_2$), 0.6 µl of $50\times$ aa-dUTP/dNTP mix (25 mM dATP, dGTP, dCTP, 10 mM dTTP and 15 mM aa-dUTP), 3 µl of 0.1 M DTT, 3 µl of 50 U/µl StrataScript™ RT (Stratagene catalog no. 600085) and 5 µl of dH$_2$O is mixed to a final volume of 30 µl. The reaction is incubated at 42 °C for 2 hrs. Ten µl of 1 N NaOH and 10 µl of 0.5 M EDTA is added to the reaction and incubated at 65 °C for 15 min to hydrolyze the RNA. Following the hydrolysis, the mixture is neutralized with the addition of 25 µl of 1 M Tris-HCl (pH 7.4) and stored at 4 °C. All of the Tris base in the reaction mixture must be removed before continuing with the amino-allyl dye coupling procedure to prevent the monofunctional NHS-ester Cy dyes from coupling to free amine groups in solution. To accomplish this, the neutralized reaction is added to a Microcon-30 concentrator (Millipore catalog no. 42422) filled with 450 µl of dH$_2$O. The Microcon-30 concentrator is centrifuged at $10000\times g$ for 8 min and the flowthrough is discarded. This centrifugation procedure is repeated. The Microcon-30 concentrator is inverted and centrifuged a third time to elute the concentrated solution. This Tris clean-up procedure is repeated with a new Microcon-30 concentrator if the volume of the eluted solution is greater than 150 µl. The eluted solution is dried in a speed-vac and stored at -20 °C. For cDNA labeling with Cy3 or Cy5, the amino-allyl cDNA pellet is resuspended in 9 µl of 0.05 M sodium bicarbonate buffer (pH 9.0) and placed at room temperature for 10–15 min to ensure resuspension. A fresh tube of monofunctional NHS-ester Cy3 or Cy5 is resuspended in 10 µl of DMSO and aliquoted into 1.25 µl $\times$ 8 tubes and immediately dried in a speed-vac. The amino-allyl cDNA sodium bicarbonate buffer mixture is transferred to the Cy3- or Cy5-containing tubes and incubated at room temperature for 1 hr in the dark. Before Cy3- and Cy5-labeled cDNA is combined for hybridization, the labeled cDNA is quenched to prevent cross-coupling by adding 4.5 μl of 4 M hydroxylamine to the Cy-dye-labeled cDNAs and incubating at room temperature for 15 min in the dark. The unincorporated Cy dyes are removed with a Qiagen-Quick PCR purification kit (Qiagen catalog no. 28104). The eluate volume is decreased to 15 μl in a Microcon-30 concentrator.

Cy3, Cy5 target labeling of poly(A) mRNA for hybridization to glass slide arrays

Although we prefer the method described above (see Chapter 4), this method can be modified for labeling poly(A) mRNA targets for hybridization to pre-synthesized full-length ORF glass slide arrays by using oligo(dT) primers instead of random hexamer primers or a mixture of oligo(dT) and random hexamer primers (5μg/μl each).

Hybridization of Cy3-, Cy5-labeled targets to glass slide arrays

One glass slide array is put into a 50-ml conical tube and filled with prehybridization buffer ($5 \times$ SSC ($20 \times$ SSC contains 3 M NaCl and 0.3 M sodium citrate), 0.1% SDS, 1% bovine serum albumin) and incubated at 65 °C in a water bath for at least 1 hr. The prehybridization buffer is removed by dipping the slide array five times in Milli-Q grade water at room temperature. The water is removed by dipping the slide array in isopropanol at room temperature and air-drying. At this point, it is important to use the slide array within 1 hr since hybridization efficiency decreases rapidly.

Next 1.5 μl of salmon sperm DNA (10 mg/ml), 3.0 μl of $20 \times$ SSC and 0.45 μl of 10% SDS are added in to 15 μl of Cy3-, Cy5-labeled cDNA target mix prepared as described above. The target mixture is heated at 95–100 °C for 3 min and centrifuged in microfuge at maximum speed for 1 min. At this point one of several methods of applying the hybridization solution to the array without disturbing it can be used. We prefer to place a glass coverslip on a flat surface and pipette 15–20 ml of hybridization onto the top of the coverslip. The glass slide array is then inverted and carefully lowered onto the coverslip. The slide is inverted again and placed in a sealed hybridization chamber (Corning Inc. catalog no. 2551) and 10 μl of dH$_2$O or $3 \times$ SSC is added to each end of chamber to prevent the glass slide array from drying out. The sealed chamber is placed in a 65 °C water bath and kept level. The chamber containing the glass slide array is incubated in the dark for 12–18 hrs. Following hybridization, the glass slide array is removed from the hybridization chamber without disturbing the coverslip. The glass slide array is placed in 50-ml conical tube containing $0.5 \times$ SSC and 0.2% SDS. The coverslip is then removed gently while the slide array is merged in solution and agitated for 3 min at room temperature. The slide array is removed and placed into a new 50-ml conical tube containing fresh wash buffer ($0.5 \times$ SSC, 0.2% SDS) and agitated for another 20 min at 65 °C. This procedure is repeated two more times. The slide array is washed with $0.1 \times$ SSC at room temperature for 4 min to remove SDS. At this point, the glass slide array is air-dried in the dark and ready for scanning.

mRNA enrichment and biotin labeling methods for hybridization of bacterial targets to *in situ* synthesized Affymetrix GeneChips™ or glass slide arrays containing oligonucleotide probes

To enrich the proportion of mRNA in a total RNA preparation, 300 μg of total RNA is split into 12 aliquots to increase the efficiency of the enrichment procedure. All reactions are performed in PCR tubes in a thermocycler. For each reaction, 25 μg of total RNA is mixed with 70 pmol of a rRNA specific primer mix in a final volume of 30 μl. Each specific primer mix includes three specific primers for 16S rRNA (5′-CCTACGGTTACCTTGTT-3′, 5′-TTAACCTTGCGGCC GTACTC-3′, and 5′-TCCGATTAACGCTTGCACCC-3′) and five specific primers for 23S rRNA (5′-CCTCACGGTTCATTAGT-3′, 5′-CTATAGTAAAGGTTCACGGG-3′, 5′-TCGTCATCACGCCTCAGCCT-3′, 5′-TCCCACATCGTTTCCCAC-3′ and 5′-CATGGAAAACATATTACC-3′). The procedure described here for *Escherichia coli* can be used for other organisms with appropriate ribosomal specific oligonucleotide primers. This mixture is heated to 70 °C for 5 min and cooled to 4°C. Now 10 μl of 10× MMLV RT buffer (0.5 M Tris-HCl (pH 8.3), 0.1 M MgCl$_2$, and 0.75 M KCl), 5 μl of 10 mM DTT, 2 μl of 25 mM dNTPs mix, 3.5 μl of 20 U/μl of SUPERase•In™ (Ambion Inc. catalog no. 2694), 6 μl of 50 U/μl of MMLV reverse transcriptase and water are added to each tube to a final volume of 100 μl. The reactions are incubated at 42 °C for 25 min and incubation is continued at 45 °C for 20 min for cDNA synthesis. To remove the rRNA moiety from the rRNA/cDNA hybrid, 5 μl of 10 U/μl of RNase H is added and the mixture is incubated at 37 °C for 45 min. RNase H is inactivated by heating at 65 °C for 5 min. Newly synthesized cDNA is removed by incubation with 4 μl of 2 U/μl of DNase I and 1.2 μl of 20 U/μl of SUPERase•In™ at 37 °C for 2 hrs. Four reactions are combined for RNA clean-up with a single Qiagen RNeasy mini column (Qiagen catalog no. 74104). The quantity of enriched mRNA is measured by absorbance at 260 nm. A typical yield is 10–20 μg of RNA from 300 μg of total RNA constituting a five- to tenfold enrichment of mRNA to rRNA.

For the RNA fragmentation step, a maximum of 20 μg of RNA is added to a PCR tube containing 10 μl of 10× NEB buffer (0.7 M Tris-HCl (pH 7.6), 0.1 M MgCl$_2$, 50 mM DTT) for T4 polynucleotide kinase in a final volume of 88 μl. The tube is incubated at 95 °C for 30 min and cooled to 4 °C.

For the RNA 5′-thiolation and biotin-labeling reaction, 2 μl of 5 mM γ-S-ATP and 10 μl of 10 U/μl of T4 polynucleotide kinase are incubated with the fragmented RNA at 37 °C for 50 min. The reaction is inactivated by heating to 65 °C for 10 min and cooled to 4 °C. Excess γ-S-ATP is removed by ethanol precipitation. Fragmented and thiolated RNA is collected by centrifugation in the presence of glycogen (0.25 μg/μl) and resuspended in 90 μl of dH$_2$O. Six μl of 500 mM MOPS (3-(N-morpholino)propanesulfonic acid, pH 7.5) and 4.0 μl of 50 mM PEO-idoacetyl-biotin (Pierce Chemical catalog no. P/N 21334ZZ) is added to the fragmented thiolated RNA and incubated at 37 °C for 1 hr. The biotin-labeled RNA is isolated by ethanol precipitation, washed twice with 70% ethanol, dried and dissolved in 20–30 μl of Milli-Q grade water. The quantity of the biotin-labeled RNA is measured by absorbance at 260 nm. The total yield for the entire procedure is typically 2–4 μg of biotin-labeled RNA from 300 μg of total RNA. The efficiency of RNA fragmentation and biotin labeling is monitored with a gel shift assay where the biotin-labeled RNA is pre-incubated with avidin prior to electrophoresis. Biotin-labeled RNAs are retarded during electrophoresis due to the avidin–biotin interaction. The position of the RNA in the gel addresses the fragmentation efficiency. The amount of shifted RNA indicates the efficiency of the biotin labeling. Inefficiencies in either of these parameters should be addressed before proceeding to the hybridization step.

Hybridization of biotinylated mRNA targets to Affymetrix GeneChips™

For hybridization of biotinylated mRNA targets to the Affymetrix GeneChips™, 2–4 μg of fragmented, biotin-labeled RNA is used for each GeneChip™. The hybridization solution for each chip is prepared with 100 μl of 2× MES hybridization buffer (200 mM 2-(N-morpholino)ethanesulfonic acid (MES), 2 M NaCl, 40 mM EDTA, and 0.02% Tween 20 detergent), 1 μl of 100 nmol Biotin-Oligo 948 (5'biotin-GTCAAGATGCTACCGTTCAG-3'), 2 μl of 10 mg/μl herring sperm DNA, 2 μl of 50 mg/ml bovine serum albumin and 2–4 μg of fragmented biotin-labeled RNA and brought to a final volume of 200 μl with Milli-Q grade water. The GeneChip™ arrays are equilibrated to room temperature immediately before use. The hybridization solution prepared above is added to each GeneChip™ and incubated in a GeneChip™ hybridization oven (Affymetrix) at 45 °C for 16 hrs at a rotation rate of 60 rpm.

Following hybridization, the stain and wash procedures are carried out in an Affymetrix GeneChip™ Fluidics Station 400 using the fluidics script for the particular GeneChip™ type to run the machine. Streptavidin solution mix (300 μl of 2× MES stain buffer (200 mM MES, 2 M (Na+), 0.1% Tween 20), 24 μl of 50 mg/ml bovine serum albumin, 6 μl of 1 mg/ml streptavidin and 270 μl of dH$_2$O), antibody solution (300 μl 2× MES stain buffer, 24 μl of 50 mg/ml bovine serum albumin, 6 μl of 10 mg/ml normal goat IgG, 6 μl of 0.5 mg/ml biotin anti-streptavidin and 264 μl of dH$_2$O) and SAPE solution (300 μl of 2× MES stain buffer, 24 μl of 50 mg/ml bovine serum albumin, 6 μl of 1 mg/ml streptavidin–phycoerythrin, and 270 μl of dH$_2$O) are prepared in amber tubes for the staining of each probe array. After hybridization, the hybridization solution is removed and kept at 4 °C. Each GeneChip™ is filled with 300 μl of non-stringent wash buffer (6× SSPE, 0.01% Tween 20, 0.005% Antifoam). The GeneChips™ are inserted into the fluidics station and the ProGE-WS2 protocol is selected from the computer drop-down menu to control the staining and washing of the probe arrays and the instructions on the LCD screen of the fluidics station are followed. Following the staining and washing procedures the GeneChips™ are removed from the fluidics station and checked for large bubbles or air pockets before scanning. The buffer in the GeneChips™ is drained and refilled with non-stringent buffer if bubbles are present.

Biotin labeling methods for hybridization of eukaryotic targets to Affymetrix GeneChips™

Typically, 10 μg of total RNA is used for labeled targets prepared with a Gibco BRL SuperScript Choice system (Gibco catalog Series 18090). In this case, 1 μl of T7-(dT)$_{24}$ primer (100 pmol/μl) and 10 μg of total RNA is mixed in a volume of 11 μl. The mix is heated to 70 °C for 10 min and put on ice after microcentrifugation for a few seconds. Four μl of 5× first strand cDNA buffer (250 mM Tris-HCl (pH 8.3), 375 mM KCl, 15 mM MgCl$_2$), 2 μl of 0.1 M DTT and 1 μl of 10 mM dNTPs mix is added to the centrifuged mixture and heated to 42 °C for 2 min. Two μl of SuperScript II RT (200 U/μl) is added and incubated at 42 °C for 1 hr. Following first-strand cDNA synthesis, 30 μl of 5× second-strand reaction buffer (100 mM

Tris-HCl (pH 6.9), 450 mM KCl, 23 mM MgCl$_2$, 750 μM β-NAD, 50 mM (NH$_4$)$_2$SO$_4$), 3 μl of 10 mM of dNTPs mix, 1 μl of DNA ligase (10 U/μl), 4 μl of DNA polymerase I (10 U/μl), 1 μl of RNase H (2 U/μl), and 91 μl of dH$_2$O is added to produce a final volume of 150 μl. This mixture is incubated at 16 °C for 2 hrs. Two μl of T4 DNA polymerase (10 U) is added and incubated at 16 °C for 5 min. Ten μl of 0.5 M EDTA is added to stop the reaction. Following the second-strand cDNA synthesis, 162 μl of (25:24:1) phenol:chloroform:isoamyl alcohol (saturated with 10 mM Tris-HCl (pH 8.0)/1 mM EDTA) is added to the reaction and vortexed. Phase Lock Gel (PLG; Brinkmann Instruments Inc. Eppendorf®, catalog no. 0032 007.961) is used to form an inert sealed barrier between the aqueous and organic phases of the phenol–chloroform extraction to allow more complete recovery of the sample (aqueous phase) and minimize interface contamination. PLG is sold in pre-measured aliquots in 1.5-ml tubes. Each PLG-containing tube is microcentrifuged at 12000× *g* or higher for 20–30 s. The vortexed phenol–chloroform sample is added to the tube containing the compressed gel. The sample is mixed with a pipette and the PLG tube is microfuged at 12000 × *g* or greater for 2 min. The aqueous phase is transferred to a fresh 1.5-ml tube. Next, 0.5 volume of 7.5 M NH$_4$Ac and 2.5 volume of ethanol (−20 °C) are added to the sample and the tube is vortexed. The mixed sample then is centrifuged immediately at greater than 12000 × *g* in a micro-centrifuge at room temperature for 20 min. The supernatant is removed and the pellet is washed with 0.5 volume of 80% ethanol (−20 °C). The sample again is centrifuged at greater than 12000 × *g* at room temperature for 5 min. The 80% ethanol is carefully removed and the 80% ethanol wash procedure is repeated one more time. The pellet is air-dried and resuspended in 12 μl of RNase-free water.

The biotin-labeled antisense cRNA (targets) for hybridization to the GeneChips™ are prepared from the double-stranded cDNA preparation described above using an *in vitro* transcription reaction (IVT) kit supplied by ENZO (BioArray HighYield™ RNA Transcript Labeling Kit, part no. 900182). Using this kit, 4 μl of 10× high yield reaction buffer, 4 μl of 10× biotin-labeled ribonu-cleotides (ATP, GTP, CTP, UTP, bio-UTP and bio-CTP), 4 μl of 10× DTT, 4 μl of 10× RNase inhibitor mix, 2 μl of 20× T7 RNA polymerase and up to 1 μg of cDNA are mixed in a RNase-free tube and the final volume is brought to 40 μl with water. The tube is placed in a 37 °C water bath and incubated for 4–5 hrs. Usually 50–100 μg of RNA product is produced in each standard reaction (40 μl reaction containing 1 μg of template DNA). It is essential to remove unincorporated NTPs before proceeding for hybridization. The quantity of cRNA can be determined by 260 nm absorbance. An aliquot of the unpurified IVT product is saved for gel electrophoresis analysis. If the total RNA used for the cDNA reaction exceeds 20 μg, it is recommended to split the IVT reaction for clean-up. RNeasy spin columns from Qiagen are used for RNA clean-up. Following instructions in the Qiagen handbook the sample is passed over the column two times before the wash and elution steps to increase yield: 50 μl of water is added to the column for RNA elution and the column is allowed to sit for 1 min before centrifugation. The elution step is repeated one more time and both eluted samples are combined. Then 0.5 volumes of 7.5 M NH$_4$Ac, 2.5 volumes of absolute ethanol and 2 μl of glycogen (5 μg/μl) are added to the elution and the sample is precipitated at −20 °C for 1 hr to overnight. The sample is collected by centrifugation at greater than 12000× *g* at 4 °C for 30 min. The pellet is washed twice with 0.5 ml of 80% ethanol (−20 °C) and air-dried. The pellet is resuspended in 20 μl of RNase-free water and checked with a spectropho-tometer at 260 nm and 280 nm absorption to determine sample concentration and purity (an A$_{260}$/A$_{280}$ ratio should be close to 2.0 for pure RNA).

For quantitation of cRNA when using total RNA as starting material, an adjusted cRNA yield must be calculated to reflect carryover of unlabeled total RNA. The adjusted cRNA yield = RNA_m – (total RNA_i)(y), where RNA_m is the amount of cRNA measured after IVT (μg), total RNA_i is the initial amount of total RNA (μg), and y is the fraction of cDNA reaction used in the IVT. Twenty μg of the adjusted yield of cRNA and 4 μl of 5× fragmentation buffer (200 mM Tris-acetate (pH 8.1), 500 mM KOAc, 150 mM MgOAc) are mixed and brought to a volume of 20 μl with RNase-free water for the fragmentation procedure. The mixture is incubated at 94 °C for 35 min. The efficiency of fragmentation is checked with gel electrophoresis in 1% TAE agarose gel. The standard fragmentation procedure should produce a distribution of RNA fragment sizes from approximately 35 to 200 bases. The fragmented sample is then stored at −20 °C until ready to perform the hybridization.

For the hybridization, 15 μg of fragmented cRNA, 3 μl of control oligonucleotide B2 (5′-biotin-GTCGTCAAGATGCTACCGTTCAGGA-3′, 5 nmol), 3 μl of 100× control cRNA cocktail (150 pmol of BioB, 500 pmol of BioC, 2.5 nmol of BioD, 10 nmol of cre, 0.1 mg/ml of herring sperm DNA, 1× MES, 0.925 M of NaCl, and 0.01% of Tween 20), 3 μl of herring sperm DNA (10 mg/ml), 3 μl of acetylated bovine serum albumin (50 mg/ml), 150 μl of 2× MES hybridization buffer and water are mixed together to a final volume of 300 μl as hybridization cocktail. GeneChips™ are equilibrated at room temperature immediately before use and filled with appropriate volume of 1× MES hybridization buffer. The GeneChip™ is incubated at 45 °C for 10 min with rotation in the hybridization oven. At the same time, the hybridization cocktail is heated to 99 °C for 5 min and then transferred to a 45 °C heat block for 5 min. The hybridization cocktail is centrifuged at maximum speed in a microcentrifuge for 5 min to remove any insoluble material from the hybridization mixture. The buffer is removed from the GeneChip™ cartridge and filled with 200 μl of hybridization cocktail prepared as above and each GeneChip™ is incubated in a GeneChip™ hybridization oven at 45 °C for 16 hrs at a rotation rate of 60 rpm.

Following hybridization, the stain and wash procedures are carried out in an Affymetrix GeneChip™ Fluidics Station 400 using the fluidics script for the particular GeneChip™ type to run the machine. Streptavidin solution mix (300 μl of 2× MES stain buffer, 24 μl of 50 mg/ml bovine serum albumin, 6 μl of 1 mg/ml streptavidin, and 270 μl of dH_2O), antibody solution (300 μl 2× MES stain buffer, 24 μl of 50 mg/ml bovine serum albumin, 6 μl of 10 mg/ml normal goat IgG, 6 μl of 0.5 mg/ml biotin anti-streptavidin, and 264 μl of dH_2O), and SAPE solution (300 μl of 2× MES stain buffer, 24 μl of 50 mg/ml bovine serum albumin, 6 μl of 1 mg/ml streptavidin–phycoerythrin, and 270 μl of dH_2O) are prepared in amber tubes for the staining of each probe array. After hybridization, the hybridization solution is removed and kept at 4 °C. Each GeneChip™ is filled with 300 μl of non-stringent wash. The GeneChips™ are inserted into the fluidics station and the EuKGE-WS2 protocol is selected from the computer drop-down menu to control the staining and washing of the probe arrays and the instructions on the LCD screen of the fluidics station are followed. Following the staining and washing procedures the GeneChips™ are removed from the fluidics station and checked for large bubbles or air pockets before scanning. The buffer in the GeneChips™ is drained and refilled with non-stringent buffer if bubbles are present.

Appendix B
Mathematical complements

In this Appendix we provide a few mathematical complements to Chapters 5 and 6. In particular, we describe the probability distributions that are used in Chapter 5 and briefly review Gaussian processes and support vector Machines (SVMs) which are mentioned in Chapter 6. We have selected SVMs over many other classification/regression techniques that could not be included here for lack of space because they were developed more recently and are less widely known.

Distributions

In Chapter 5, we have modeled the expression level of a gene under a given condition using a Gaussian distribution parameterized by its mean and variance. We have used a conjugate prior for the mean and variance and provided arguments in support of this choice. Other prior distributions for the parameters of a Gaussian distribution are possible and, for completeness, here we describe two such alternatives which are studied in the literature [1, 2]. We leave as an exercise for the reader to derive the corresponding estimates and compare them to those derived in Chapter 5.

The non-informative improper prior

The standard non-informative but improper prior for μ and σ^2 is to assume that they are independent of each other and uniformly distributed. More precisely, since σ^2 is a scale parameter, this prior is a uniform prior for μ and $\log \sigma$. $P(\mu, \log \sigma) = C$ or, equivalently, $P(\mu, \sigma^2) = C/\sigma^2$. This prior is improper because it does not integrate to 1. A simple calculation shows that the proper posterior is:

$$P(\mu, \sigma^2 | D) = C\sigma^{-(n+2)} \exp\left[-\frac{n-1}{2\sigma^2} s^2 - \frac{n}{2\sigma^2} (\mu - m)^2 \right] \qquad (B.1)$$

with $-\infty < \mu < +\infty$ and $\sigma > 0$. The conditional and marginal posteriors are easily derived. The conditional posterior $P(\mu | \sigma, D)$ is normal $\mathcal{N}(m, \sigma^2/n)$. The marginal posterior of μ is a Student $t(n-1, m, s^2/n)$ distribution. The marginal posterior distribution of σ^2 is a scaled inverse gamma density $P(\sigma^2 | D) = I(\sigma^2; n-1, s^2)$.

A better prior in the case of gene expression data should assume that $\log \sigma$ is uniform over a finite interval $[a, b]$ or that σ^2 has a bell-shaped distribution concentrated on positive values, such as a gamma distribution. The conjugate prior addresses these issues.

The semi-conjugate prior

In the semi-conjugate prior distribution, the functional form of the prior distributions on μ and σ^2 are the same as in the conjugate case (normal and scaled inverse gamma, respectively) but independent of each other, i.e., $P(\mu, \sigma^2) = P(\mu)P(\sigma^2)$. However, as previously discussed, this assumption is not optimal for current DNA array data.

Other more complex priors could be constructed using mixtures. A mixture of conjugate priors would lead to a mixture of conjugate posteriors.

The scaled inverse gamma distribution

The scaled inverse gamma distribution $I(x; v, s^2)$ with $v > 0$ degrees of freedom and scale $s > 0$ is given by:

$$\frac{(v/2)^{v/2}}{\Gamma(v/2)} s^v x^{-(v/2+1)} e^{-vs^2/(2x)} \tag{B.2}$$

for $x > 0$. Γ represents the gamma function $\Gamma(y) = \int_0^\infty e^{-t} t^{y-1} dt$. The expectation is $(v/v-2)s^2$ when $v > 2$; otherwise it is infinite. The mode is always $(v/v+2)s^2$.

The Student t distribution

The Student t distribution $t(x; v, m, \sigma^2)$ with $v > 0$ degrees of freedom, location m, and scale $\sigma > 0$ is given by:

$$\frac{\Gamma((v+1)/2)}{\Gamma(v/2)\sqrt{v\pi\sigma}} \left[1 + \frac{1}{v} \left(\frac{x-m}{\sigma} \right)^2 \right]^{-(v+1)/2} \tag{B.3}$$

The mean and the mode are equal to m.

The inverse Wishart distribution

The inverse Wishart distribution $I(W; v, S^{-1})$, where v represents the degrees of freedom and S is a $k \times k$ symmetric, positive definite scale matrix, is given by:

$$\left[2^{vk/2} \pi^{k(k-1)/4} \prod_{i=1}^{k} \Gamma\left(\frac{v+1-i}{2} \right) \right]^{-1} |S|^{v/2} |W|^{-(v+k+1)/2}$$

$$\exp\left(-\frac{1}{2} tr(SW^{-1}) \right) \tag{B.4}$$

where W is also positive definite. The expectation of W is $E(W) = (v-k-1)^{-1}S$.

The beta distribution

The beta distribution $B(x; r, s)$, also called two-dimensional Dirichlet distribution, with parameters $r > 0$ and $s > 0$, is defined by

$$B(x; r, s) = \frac{\Gamma(r+s)}{\Gamma(r)\Gamma(s)} x^{r-1}(1-x)^{s-1} \tag{B.5}$$

with $0 \le x \le 1$. The mean is $r/(r+s)$ and the variance $rs/(r+s)^2)(r+s+1)$. It is a useful distribution for quantities that are constrained to the unit interval, such as probabilities. When $r=s=1$ it yields the uniform distribution. For most other values, it yields a "bell-shaped" distribution over the [0, 1] interval.

Hypothesis testing

We have modeled the log-expression level of each gene in each situation using a Gaussian model. If all we care about is whether a given gene has changed or not, we could model directly the difference between the log-expression levels in the control and treatment cases. These differences can be considered pairwise or in paired fashion, as is more likely the case with current microarray technology where the logarithm of the ratio between the expression levels in the treatment and control situations is measured along two different channels (red and green).

We can model again the differences $x^t - x^c$ with a Gaussian $\mathcal{N}(\mu, \sigma^2)$. Then the null hypothesis H, given the data, is that $\mu = 0$ (no change). To avoid assigning a probability of 0 to the null hypothesis, a Bayesian approach here must begin by giving a non-zero prior probability for $\mu = 0$, which is somewhat contrived. In any case, as in Chapter 5, we can set $P(\sigma^2) = I(\sigma^2; v_0, \sigma_0^2)$. For the mean μ, we use the mixture

$$\mu = \begin{cases} 0 & : \text{ with probability } p \\ \mathcal{N}(0, \sigma^2/\lambda) & : \text{ with probability } 1-p \end{cases} \tag{B.6}$$

The parameter p could be fixed from previous experiments, or treated as an hyperparameter with, for instance, a Dirichlet prior. We leave as an exercise for the reader to compute the relevant statistics $\log [P(\bar{H})/P(H)]$.

Missing values

It is not uncommon for array data to have missing values resulting, for instance, from experimental errors. Ideally, missing values should be dealt with probabilistically by estimating their distribution and integrating them out. When they are not removed entirely from the analysis, a more straightforward approximation is to replace missing values by single point estimates. At the crudest level, a missing level of expression could be replaced by the average of the expression levels of all the genes in the same experiment. More adequate estimates can be derived using the techniques of Chapter 5, by looking at "neighboring" genes with similar properties (see also [3]).

Gaussian process models

Consider a regression problem consisting of K input–output training pairs (x_1, y_1), ..., (x_K, y_K) drawn from some unknown distribution. The inputs x are n-dimensional vectors. For simplicity, we assume that y is one-dimensional, but the extension to the multi-dimensional case is straightforward. xs could represent the level of activity of a set of genes, and y some physiological measurement (including the activity of a gene not contained in the set) or, in a classification problem, a categorical prediction such as cancer/non-cancer. The goal in regression is to learn the functional relationship between x and y, from the given examples. The Gaussian process modeling approach [4, 5, 6], also known as "kriging", provides a flexible probabilistic framework for regression and classification problems. In fact a number of non-parametric regression models, including neural networks with a single infinite hidden layer and Gaussian weight priors, are equivalent to Gaussian processes [7]. Gaussian processes, however, can be used to define probability distributions over spaces of functions directly, without any need for an underlying neural architecture.

A Gaussian process is a collection of variables $Y = [y(x_1), y(x_2), \ldots]$, with a joint Gaussian distribution of the form

$$P(Y|C, \{x_i\}) = \frac{1}{Z} \exp\left(-\frac{1}{2}(Y - \mu)^T C^{-1}(Y - \mu) \right) \tag{B.7}$$

for any sequence $\{x_i\}$, where μ is the mean vector and $C_{ij} = C(x_i, x_j)$ is the covariance of x_i and x_j. For simplicity, we shall assume in what follows that $\mu = 0$. Priors on the noise and the modeling function are combined into the covariance matrix C. Different sensible parameterizations for C are described below. From Equation B.7, the predictive distribution for the variable y associated with a test case x, is obtained by conditioning on the observed training examples. In other words, a simple calculation shows that y has a Gaussian distribution

$$P(y|\{y_1, \ldots, y_K\}, C(x_i, x_j), \{x_1, \ldots, x_K, x\}) = \frac{1}{\sqrt{2\pi}\sigma} \exp\left(-\frac{(y - y^*)^2}{2\sigma^2} \right) \tag{B.8}$$

with

$$y^* = k(x)^T C_K^{-1}(y_1, \ldots, y_K) \quad \text{and} \quad \sigma = C(x, x) - k(x)^T C_K^{-1} k(x) \tag{B.9}$$

where $k(x) = (C(x_1, x), \ldots, C(x_K, x))$, and C_K denotes the covariance matrix based on the K training samples.

Covariance parameterization

A Gaussian process model is defined by its covariance function. The only constraint on the covariance function $C(x_i, x_j)$ is that it should yield positive semi-definite matrices for any input sample. In the stationary case, Bochner theorem in harmonic analysis [8] provides a complete characterization of such functions in terms of Fourier transforms. It is well known that the sum of two positive matrices (resp. positive definite) is positive (resp. positive definite). Therefore the covariance can be conveniently parameterized as a sum of different positive components. Examples of useful components have the following forms:

- Noise variance: $\delta_{ij}\theta_1^2$ or, more generally, $\delta_{ij} f(x_i)$ for an input dependent-noise model where $\delta_{ij} = 0$ if $i \neq j$ and $\delta_{ij} = 1$ otherwise
- Smooth covariance: $C(x_i, x_j) = \theta_2^2 \exp(-\sum_{u=1}^{n} \rho_u^2 (x_{iu} - x_{ju})^2)$
- And more generally: $C(x_i, x_j) = \theta_2^2 \exp(-\sum_{u=1}^{n} \rho_u^2 |x_{iu} - x_{ju}|^r)$
- Periodic covariance: $C(x_i, x_j) = \theta_3^2 \exp(-\sum_{u=1}^{n} = \rho_u^2 \sin^2[\pi(x_{iu} - x_{ju})/\gamma_u]$

Notice that a small value of ρ_u characterizes components u that are largely irrelevant for the output in a way closely related to the automatic relevance determination framework [7]. For simplicity, we write θ to denote the vector of hyperparameters of the model. Short of conducting lengthy Monte Carlo integrations over the space of hyperparameters, a single value θ can be estimated by minimizing the negative log-likelihood

$$\varepsilon(\theta) = \frac{1}{2}\log \det C_K + \frac{1}{2} Y_K^T C_K^{-1} Y_K + \frac{K}{2}\log 2\pi \tag{B.10}$$

Without any specific shortcuts, this requires inverting the covariance matrix and is likely to require $O(N^3)$ computations. Prediction or classification can then be carried based on Equation B.9. A binary classification model, for instance is readily obtained by defining a Gaussian process on a latent variable Z as above and letting

$$P(y_i = 1) = \frac{1}{1 + e^{-z_i}} \tag{B.11}$$

More generally, when there are more than two classes, one can use normalized exponentials instead of sigmoidal functions.

Kernal methods and support vector machines

Kernal methods and Support Vector Machines (SVMs) [9, 10] are related to Gaussian processes and can also be used in classification and regression problems. These techniques are particularly suited for modeling complex discrimination boundaries and have been applied to DNA array data, for instance, to discrimination between gene families or cancer types. While in the short run the sophistication of SVMs may be somewhat of an overkill, as array technology improves and array data expands, these and other related methods are likely to play a significant role.

For simplicity, here we consider a binary classification problem characterized by a set of labeled training example pairs of the form (x_i, y_i) where x_i is an input vector and $y_i = \pm 1$ is the corresponding classification in one of two classes H^+ and H^- (a (0,1) formalism is equivalent but leads to more cumbersome notations). As an example, the reader may consider the problem of deciding whether a given gene belongs to a given family given the expression levels of members within (positive examples) and outside the family (negative examples) [11, 12]. It is possible for the length of x_i to vary with i. The label y for a new example x is determined by a discriminant function $\mathcal{D}(x; \{x_i, y_i\})$, which depends on the training examples, in the form $y = \text{sign}(\mathcal{D}(x; \{x_i, y_i\}))$. In a proper probabilistic setting,

$$y = \text{sign}(\mathcal{D}(x; \{x_i, y_i\})) = \text{sign}\left(\log \frac{P(H^+|x)}{P(H^-|x)}\right) \tag{B.12}$$

In kernel methods, the discriminant function is expanded in the form

$$\mathcal{D}(x) = \sum_i y_i \lambda_i K(x_i, x) = \sum_{H^+} \lambda_i K(x_i, x) - \sum_{H^-} \lambda_i K(x_i, x) \qquad \text{(B.13)}$$

so that, up to trivial constants, $\log P(H^+|x) = \sum_{H^+} \lambda_i K(x_i, x)$ and similarly for the negative examples. K is called the kernel function. The intuitive idea is to base the classification of the new examples on all the previous examples weighted by two factors: a coefficient $\lambda_i \geq 0$ measuring the importance of example i, and the kernel $K(x_i, x)$ measuring how similar x is to example x_i. Thus in a sense the expression for the discrimination depends *directly* on the training examples. This is different from the case of neural networks, for instance, where the decision depends indirectly on the training examples via the trained neural network parameters. Thus in an application of kernel methods two fundamental choices must be made regarding the kernel K and the weights λ_i. Variations in these choices lead to a spectrum of different methods, including generalized linear models and SVMs.

Kernel selection

To a first approximation, from the mathematical theory of kernels, a kernel must be positive definite. By Mercer's theorem of functional analysis, K can be represented as an inner product of the form

$$K_{ij} = K(x_i, x_j) = \phi(x_i)\phi(x_j) \qquad \text{(B.14)}$$

Thus another way of looking at kernel methods, is to consider that the original x vectors are mapped to a "feature" space via the function $\phi(x)$. Note that the feature space can have very high (even infinite) dimension and that the vectors $\phi(x)$ have the same length even when the input vectors x do not. The similarity of two vectors is assessed by taking their inner production in feature space. In fact we can compute the Euclidean distance $\|\phi(x_i) - \phi(x_j)\|^2 = K_{ii} - 2K_{ij} + K_{jj}$ which also defines a pseudo-distance on the original vectors.

The fundamental idea in kernel methods is to define a linear or non-linear decision surface in feature space rather than the original space. The feature space does not need to be constructed explicitly since all decisions can be made through the kernel and the training examples. In addition, as we are about to see, the decision surface depends *directly* on a *subset* of the training examples, the support vectors.

Notice than a dot product kernel provides a way of comparing vectors in feature space. When used directly in the discrimination function, it corresponds to looking for linear separating hyperplanes in feature space. However more complex decision boundaries in feature spaces (quadratic or higher order) can easily be implemented using more complex kernels K' derived from the inner product kernel K, such as:

- Polynomial kernels: $K'(x_i, x_j) = (1 + K(x_i, x_j))^m$.
- Radial basis kernels: $K'(x_i, x_j) = \exp -\frac{1}{2\sigma^2}(\phi(x_i) - \phi(x_j))^t(\phi(x_i) - \phi(x_j))$.
- Neural network kernels: $K'(x_i, x_j) = \tanh(\mu x_i^t x_j + \kappa)$.

Another important class of kernels that can be derived from probabilistic generative models are Fischer kernels [11, 13].

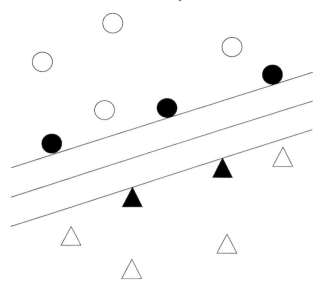

Figure B.1. Hyperplane separating two classes of points with the corresponding margins and support vectors (black circles and triangles).

Weight selection

The weights λ_i are typically obtained by iterative optimization of an objective function (classification loss), corresponding in general to a quadratic optimization problem. With large training sets, at the optimum many of the weights are equal to 0. The only training vectors that matter in a given decision are those with non-zero weights and these are called the support vectors (Figure B.1).

To see this, consider an example x_i with target classification y_i. Since the decision is based on the sign of $\mathcal{D}(x_i)$, ideally we would like to have $y_i \mathcal{D}(x_i)$, the margin for example i, to be as large possible. Because the margin can be rescaled by rescaling the λs, it is natural to introduce additional constraints such as $0 \le \lambda_i \le 1$ for every λ_i. In the case where an exact separating manifold exists in feature space, then a reasonable criterion is to maximize the margin in the worst case. This is also called risk maximization and corresponds to: $\max_\lambda \min_i y_i \mathcal{D}(x_i)$. SVMs can be defined as a class of kernel methods based on structural risk minimization [9, 10, 14]. Substituting the expression for D in terms of the kernel yields: $\max_\lambda \min_i \Sigma_j \lambda_j y_i y_j K_{ij}$. This can be rewritten as: $\max_\lambda \min_i \Sigma_j A_{ij} \lambda_j$, with $A_{ij} = y_i y_j K_{ij}$ and $0 \le \lambda_i \le 1$. It is clear that in each minimization procedure all weights λ_j associated with a non-zero coefficient A_{ij} will either be 0 or 1. With a large training set many of them will be zero for each i and this will remain true at the optimum. When the margins are violated, as in most real-life examples, we can maximize the average margin, the average being taken with respect to the weights λ_i themselves which are intended to reflect the relevance of each example. Thus in general we want to maximize a quadratic expression of the form $\Sigma_i \lambda_i y_i \mathcal{D}(x_i)$ under a set of linear constraints on the λ_i. Standard techniques exist to carry such optimization. For example a typical function used for minimization in the literature is:

Appendix B

$$\varepsilon(\lambda_i) = -\sum_i [y_i \lambda_i \mathcal{D}(x_i) + 2\lambda_i] \tag{B.15}$$

The solution to this constrained optimization problem is unique provided that for any finite set of examples the corresponding kernel matrix K_{ij} is positive definite. It can be found with standard iterative methods, although the convergence can sometimes be slow. To accommodate training errors or biases in the training set, the kernel matrix K can be replaced by $K + \mu D$, where D is a diagonal matrix whose entries are either d^+ or d^- in location corresponding to positive and negative examples [9, 10, 13, 14]. An example of application of SVMs to gene expression data can be found in [12].

In summary, kernel methods and SVMs have several attractive features. As presented, these are supervised learning methods that can leverage labeled data. These methods can build flexible decision surfaces in high dimensional feature spaces. The flexibility is related to the flexibility in the choice of the kernel function. Overfitting can be controlled through some form of margin maximization. These methods can handle inputs of variable lengths such as biological sequences, as well as large feature spaces. Feature spaces need not be constructed explicitly since the decision surface is entirely defined in terms of the kernel function and typically a sparse subset of relevant training examples, the support vectors. Learning is typically achieved through iterative solution of a linearly constrained quadratic optimization problem.

REFERENCES

1. Box, G. E. P., and Tiao, G. C. *Bayesian Inference in Statistical Analysis*. 1973. Addison Wesley.
2. Pratt, J. W., Raiffa, H., and Schlaifer, R. *Introduction to Statistical Decision Theory*. 1995. MIT Press, Cambridge, MA.
3. Troyanskaya, Cantor, M., Sherlock, G., Brown, P., Hastie, T., Tibshirani, R., Botstein, D., and Altman, R. Missing value estimation methods for DNA microarrays. 2001. *Bioinformatics* 17:520–525.
4. Williams, C. K. I., and Rasmussen, C. E. Gaussian processes for regression. In D. S. Touretzky, M. C. Mozer, and M. E. Hasselmo, editors, *Advances in Neural Information Processing Systems*, vol. 8. 1996. MIT Press, Cambridge, MA.
5. Gibbs, M. N., and MacKay, D. J. C. Efficient implementation of Gaussian processes. 1997. Technical Report, Cavendish Laboratory, Cambridge.
6. Neal, R. M. Monte Carlo implementation of Gaussian process models for Bayesian regression and classification. 1997. Technical Report No. 9702, Department of Computer Science, University of Toronto.
7. Neal, R. M. *Bayesian Learning for Neural Networks*. 1996. Springer-Verlag, New York.
8. Feller, W. *An Introduction to Probability Theory and its Applications*, vol. 2, 2nd edn. 1971. Wiley, New York.
9. Vapnik, V. *The Nature of Statistical Learning Theory*. 1995. Springer-Verlag, New York.
10. Cristianini, N., and Shawe-Taylor, J. *An Introduction to Support Vector Machines*. 2000. Cambridge University Press, Cambridge.
11. Jaakkola, T. S., Diekhans, M., and Haussler, D. Using the Fisher kernel method to detect remote protein homologies. In T. Lengauer, R. Schneider, P. Bork, D. Brutlag, J. Glasgow, H. W. Mewes, and R. Zimmer, editors, *Proceedings of the 7th International Conference on Intelligent Systems for Molecular Biology (ISMB99)*, pp. 149–155. 1999. AAAI Press, Menlo Park, CA.
12. Brown, M. P. S., Grundy, W. N., Lin, D., Cristianini, N., Walsh Sugnet, C., Ares, M. Jr., Furey, T. S., and Haussler, D. Knowledge-based analysis of microarray gene

expression data by using support vector machines. 2000. *Proceedings of the National Academy of Sciences of the USA* 97:262–267.
13. Baldi, P., and Brunak, S. *Bioinformatics: The Machine Learning Approach*, 2nd edn. 2001. MIT Press, Cambridge, MA.
14. Burges, C. J. C. A tutorial on support vector machines for pattern recognition. 1998. *Data Mining and Knowledge Discovery* 2:121–167.

Appendix C
Internet resources

This Appendix provides a short list of Internet resources and pointers. By its very nature, it is bound to be incomplete and should serve only as a starting-point.

Tools, forums, and pointers

aMAZE Data Base and pointers
http://www.ebi.ac.uk/research/pfbp
http://www.ebi.ac.uk/research/pfbp/texts/biochemical_networks_web.html

Archives for biological software and databases
http://www.gdb.org/Dan/softsearch/biol-links.html

Array software (general)
http://www.genomics.uci.edu

Array protocols and software
http://www.microarrays.org

BioCatalog
http://www.ebi.ac.uk/biocat/e-mail_Server_ANALYSIS.html

Brazma microarray page at EBI
http://industry.ebi.ac.uk/~brazma/Data-mining/microarray.html

Brown's laboratory guide to microarraying
http://cmgm.stanford.edu/pbrown

CyberT (DNA array data analysis server)
http://www.genomics.uci.edu

DNA microarray technology to identify genes controlling spermatogenesis
http://www.mcb./arizona.edu/wardlab/microarray.html

EBI molecular biology software archive
http://www.ebi.ac.uk/software/software.html

Gene-X (array data management and analysis system)
http://www.ncgr.org/research/genex

GNA (Genetic Network Analyzer) software
http://bacillus.inrialpes.fr/gna/index.html

Matern's DNA microarray page
http://barinth.tripod.com/chips.html

MGED Group (Microarray Gene Expression Data Group)
http://www.mged.org

Microarray web page Ecole Normale Superieure (Paris)
http://www.biologie.ens.fr/en/genetiqu/puces/microarraysframe.html

NCI-LECB MAExplorer (Microarray Explorer)
http://www-lecb.ncifcrf.gov/MAExplorer

NHGRI ArrayDB
http://genome.nhgri.nih.gov/arraydb

NIEHS MAPS (Microarray Project System)
http://dir.niehs.nih.gov/microarray/software/maps

Public source for microarraying information, tools, and protocols
http://www.microarrays.org

Stanford MicroArray Forum
http://cmgm.stanford.edu/cgi-bin/cgiwrap/taebshin/dcforum/dcboard.cgi

UCSD Microarray Resources (2HAPI)
http://array.sdsc.edu

Visualization for bioinformatics
http://industry.ebi.ac.uk/~alan/VisSupp

Web resources on gene expression and DNA array technologies
http://industry.ebi.ac.uk/~alan/MicroArray

Weisshaar's listing of DNA microarray links
http://www.mpiz-koeln.mpg.de/~weisshaa/Adis/DNA-array-links.html

yMGV: an interactive on-line tool for visualization and data mining of published
genome-wide yeast expression data
http://www.biologie.ens.fr/fr/genetiqu/puces/publications/ymgv/index.html

DNA array databases

ArrayExpress: The ArrayExpress Database at EBI
http://www.ebi.ac.uk/arrayexpress

Dragon Database (Database Referencing of Array Genes ONline)
http://pevsnerlab.kennedykrieger.org/dragon.htm

Drosophila Microarray Project and Metamorphosis Time Course Database
http://quantgen.med.yale.edu

ExpressDB at Harvard (George Church)
http://arep.med.harvard.edu/ExpressDB

Maxd (Microarray Group at Manchester Bioinformatics)
http://www.bioinf.man.ac.uk/microarray/maxd/index.html

NCBI Geo (Gene Expression Omnibus)
http://www.ncbi.nlm.nih.gov/geo

RAD RNA Abundance Database (University of Pennsylvania)
http://www.cbil.upenn.edu/RAD2

SMD: Stanford Microarray Database
http://genome-www4.stanford.edu/MicroArray/SMD

Whitehead Institute chipDB (Young's laboratory)
http://young39.wi.mit.edu/chipdb_public

YMD: Yale Microarray Database
http://info.med.yale.edu/microarray

yMGV: yeast Microarray Global Viewer
http://transcriptome./ens.fr/ymgv

Commercially available technical information

Affymetrix, Inc., www.affymetrix.com; Protogene Laboratories,
www.protogene.com; Agilent Technologies, www.chem.agilent.com; Rosetta
Inpharmatics, www.rii.com; Merck & Co., Inc., www.merck.com; NimbleGen,
www.nimblegen. com; CombiMatrix, www.combimatrix.com; Nanogen,
www.nanogen.com; Motorola Life Sciences, www.motorola.com/lifesciences;
Interactiva, www.interactiva.de; Corning, www.corning.com; Sigma-Genosys,
www.genosys.com; Clontech, www.clontech.com; Research Genetics,
www.resgen.com; Lynx, www.lynxgen.com; Quantum Dot Corp.,www.qdots.com.

Appendix D

CyberT: An online program for the statistical analysis of DNA array data

CyberT is an internet-based program designed to accept data in the large data spreadsheet format that is generated as output by software typically used to analyze array experiment images. Figure D.1 shows a screen-shot of the CyberT window for analyzing control versus experimental data of the type considered in this book.

Each data element may correspond to a single measurement on the array (typical of membrane- or glass-slide-based arrays) or the result of a set of measurements (typical of Affymetrix GeneChips™). This data file is uploaded to CyberT using the "Browse" button in the CyberT browser window. Another window displays a version designed for the analysis of two-dye ratio data that is generated with glass slide arrays probed with Cy3/Cy5-labeled cDNA. Both of these CyberT applications are available for use at the UCI genomics web site (www.genomics.uci.edu). Detailed instructions for using CyberT can be accessed from this web page.

An example of the first few rows of a data file for CyberT is shown in Figure 7.4. The data file can be uploaded either as a whitespace, tab, or comma delimited text file.

After uploading the data file, the user enters experiment comments and defines the columns on which analyses will be performed. In the example shown in Figure D.2, there is only one label column containing gene names; however, any number of label columns containing other gene labels (descriptions) can be designated. Since column numbering starts with 0 (not 1), the gene names are in column 0, column numbers 1, 2, 3, and 4 contain the control data, and the columns 5, 6, 7, and 8 contain the experimental data.

The next three boxes allow the user to instruct CyberT where the data to be analyzed is located in the data file. Since only the first column of the data in Figure D.2 contains gene labels, 0 is typed into the Label Columns box. To identify the columns containing the control (Lrp^+) data, 1 2 3 4 (separated by single spaces) is typed into the Control Data box, and 5 6 7 8 is typed into the Experimental Data box to identify the column positions of the experimental (Lrp^-) data. The minimum number of non-zero (above background) replicate experiment measurements to be used for the t-test (minimum of two) from the control and experimental data for each gene is entered into the next box.

The dark blue section contains data boxes for the handling of low values. For example, the lowest value to be used can be specified. All values below this value will be set to 0 and ignored in the t-test. This feature is especially valuable for Affymetrix data that contain negative values.

The light green section contains data boxes for entering parameters for the Bayesian analysis. The first box allows the user to define the size of the window used

Genex: CyberT - Statistical Analysis for Large Scale Gene Expression Data

General Help | *Please Click to Register* (to count users for grant appl'ns and for bug notifications) | Example Control/Experimental Data Set

Data File to Upload: (Format expected, Data Coding)
If your file won't upload, or Cyber-T won't process it, check these possible reasons.

wt-nolrp(4x4)(4290).csv   Browse...

Data fields delimited by: commas ▼   * whitespace = TABS & spaces
☑ Delete lines with NULL Labels.

Please enter any text that you would like to have as a header for the analysis output.

Columns start at 0, not 1; leading / lagging spaces are bad.
Label Columns (as # # #...):
0

Control Data Replicate Columns (as # # #...):
1 2 3 4

Experimental Data Replicate Columns (as # # #...):
5 6 7 8

Minimum non-zero Replicates Required (#):
4   If left blank, the number of values of Experimental Data will be used.

Low Value Handling

◉ Values less than [0] will be set to **0** and ignored in calculations. (Including negative values will cause the app to die. If you need to correct for negatives, use the option below)

◯ Offset the values by the lowest value in the dataset, effectively right-shifing the entire dataset. Useful if you want to include negative values. This value will be noted in the output.

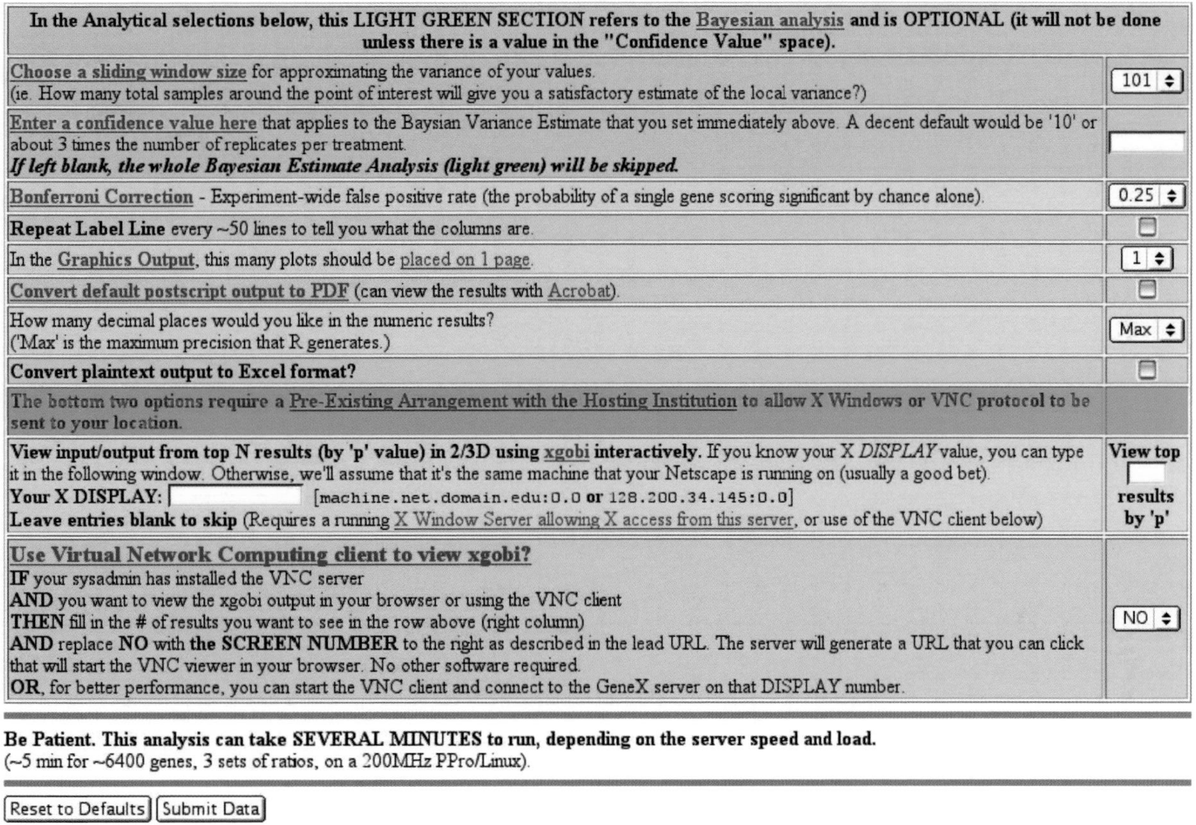

In the Analytical selections below, this **LIGHT GREEN SECTION** refers to the Bayesian analysis and is OPTIONAL (it will not be done unless there is a value in the "Confidence Value" space).

Choose a sliding window size for approximating the variance of your values. (ie. How many total samples around the point of interest will give you a satisfactory estimate of the local variance?)	101
Enter a confidence value here that applies to the Baysian Variance Estimate that you set immediately above. A decent default would be '10' or about 3 times the number of replicates per treatment. *If left blank, the whole Bayesian Estimate Analysis (light green) will be skipped.*	
Bonferroni Correction - Experiment-wide false positive rate (the probability of a single gene scoring significant by chance alone).	0.25
Repeat Label Line every ~50 lines to tell you what the columns are.	☐
In the Graphics Output, this many plots should be placed on 1 page.	1
Convert default postscript output to PDF (can view the results with Acrobat).	☐
How many decimal places would you like in the numeric results? ('Max' is the maximum precision that R generates.)	Max
Convert plaintext output to Excel format?	☐

The bottom two options require a Pre-Existing Arrangement with the Hosting Institution to allow X Windows or VNC protocol to be sent to your location.

View input/output from top N results (by 'p' value) in 2/3D using xgobi interactively. If you know your X *DISPLAY* value, you can type it in the following window. Otherwise, we'll assume that it's the same machine that your Netscape is running on (usually a good bet). **Your X DISPLAY:** [machine.net.domain.edu:0.0 **or** 128.200.34.145:0.0] Leave entries blank to skip (Requires a running X Window Server allowing X access from this server, or use of the VNC client below)	**View top** [ ] **results by 'p'**
Use Virtual Network Computing client to view xgobi? IF your sysadmn has installed the VNC server AND you want to view the xgobi output in your browser or using the VNC client THEN fill in the # of results you want to see in the row above (right column) AND replace NO with the **SCREEN NUMBER** to the right as described in the lead URL. The server will generate a URL that you can click that will start the VNC viewer in your browser. No other software required. **OR**, for better performance, you can start the VNC client and connect to the GeneX server on that DISPLAY number.	NO

Be Patient. This analysis can take SEVERAL MINUTES to run, depending on the server speed and load. (~5 min for ~6400 genes, 3 sets of ratios, on a 200MHz PPro/Linux).

[Reset to Defaults] [Submit Data]

Figure D.1. The CyberT interface at the UCI genomics web site.

Gene	C1	C2	C3	C4	E1	E2	E3	E4
aas	7.22E-05	6.44E-05	7.51E-05	6.55E-05	8.29E-05	5.70E-05	6.33E-05	3.55E-05
aat	9.80E-05	0.00E+00	0.00E+00	1.23E-06	1.00E-04	4.14E-06	0.00E+00	1.64E-04
abc	3.07E-04	2.21E-04	3.43E-04	3.73E-04	3.53E-04	2.13E-04	2.12E-04	3.30E-04
accA	1.76E-04	2.50E-04	2.52E-04	2.68E-04	2.08E-04	2.99E-04	3.44E-04	2.50E-04
accB	2.05E-04	4.50E-04	2.50E-04	3.28E-04	1.20E-04	3.78E-04	3.24E-04	2.82E-04
accC	3.71E-04	6.50E-04	4.75E-04	5.26E-04	2.68E-04	5.90E-04	5.71E-04	4.66E-04
accD	4.26E-04	2.65E-04	3.59E-04	3.99E-04	2.98E-04	2.26E-04	2.19E-04	2.34E-04
aceA	1.79E-05	0.00E+00	0.00E+00	0.00E+00	1.51E-05	0.00E+00	0.00E+00	2.44E-05
aceB	2.00E-05	0.00E+00	1.35E-05	2.39E-06	2.68E-05	1.47E-06	0.00E+00	2.01E-05
aceE	8.23E-04	8.90E-04	1.23E-03	9.70E-04	8.31E-04	9.17E-04	7.72E-04	7.62E-04
aceF	4.83E-04	4.70E-04	7.12E-04	4.88E-04	4.83E-04	4.23E-04	3.22E-04	2.64E-04
aceK	8.20E-06	0.00E+00	0.00E+00	0.00E+00	8.95E-06	0.00E+00	0.00E+00	0.00E+00
ackA	2.87E-04	3.00E-04	2.28E-04	2.78E-04	2.42E-04	2.55E-04	2.09E-04	2.45E-04
acnA	3.27E-05	1.75E-06	1.00E-05	1.54E-05	8.24E-05	2.45E-05	4.50E-05	1.03E-05
acnB	1.13E-03	8.59E-04	1.25E-03	1.36E-03	1.22E-03	7.92E-04	1.18E-03	1.17E-03
acpD	6.13E-06	2.06E-05	8.06E-06	3.43E-06	1.24E-06	1.44E-06	0.00E+00	0.00E+00
acpP	1.09E-03	1.42E-03	2.28E-03	2.30E-03	1.07E-03	2.00E-03	1.24E-03	1.02E-03
acpS	2.57E-04	4.54E-04	1.66E-04	5.12E-04	7.59E-05	3.94E-04	5.98E-04	4.72E-04
acrA	1.75E-04	2.62E-04	2.44E-04	2.26E-04	2.65E-04	3.85E-04	3.31E-04	2.97E-04
acrB	4.96E-06	0.00E+00	0.00E+00	0.00E+00	0.00E+00	0.00E+00	0.00E+00	0.00E+00

Figure D.2. An example of the format of a CyberT input data file.

to compute the background standard deviation. If a window of size w is selected, the Bayesian prior for a given gene is estimated as the average standard deviation of a window of w genes ranked by expression level, centered on the gene of interest. We have empirically determined that a window size of 101 performs well with data sets of 4000–8000 genes [1, 2]. The next data box allows the user to define an integer "confidence" given to the Bayesian prior estimate of within treatment variance. Larger confidence gives greater weight to the Bayesian prior and smaller confidence gives greater weight to the experimentally observed variance for the particular gene being tested. We have observed satisfactory performance when the actual number of experimental replicates is from two to four and the confidence value is ten. If this data box is left empty (the default setting) only the t-test analysis is performed.

The next data box in the blue section allows the user to define a Bonferroni significance level: the probability of a single gene being identified as significantly different between the control and experimental treatments by chance alone given the number of genes examined (Chapter 5). Additional boxes exist to control graphic and data output formatting (Figure 7.3). The program is executed by clicking the "Submit" button at the bottom of the web page. A complete description of the use of CyberT and a comprehensive help manual can be accessed by the "General Help" link at the top of the web page.

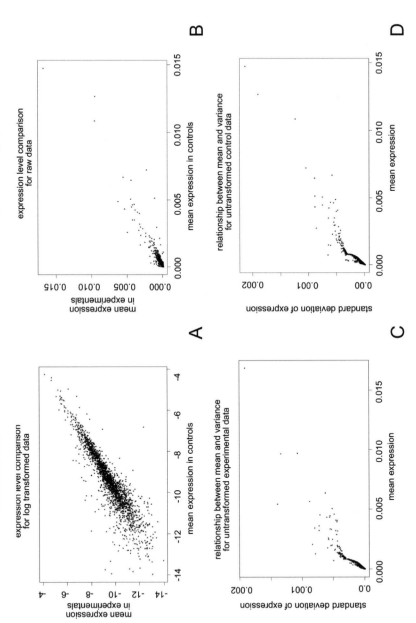

Figure D.3. An example of a CyberT graphics output file. The data presented is described in Arfin *et al.* [3]. In all cases "Experimental" refers to IHF⁻ *E. coli* cells and "Controls" refers to IHF⁺ *E. coli* cells. Panels B, C, and D are analyses on raw data, whereas the remainder of the panels are analyses carried out on log-transformed data. Log transformations of raw data generally change both average expression level and the standard deviation in expression over replicates. Panels E and F incorporate the Bayesian approach described in the text to stabilize estimates of the within-gene standard deviation, whereas panels G and H do not incorporate a Bayesian prior. The output is essentially that generated automatically by the CyberT program, except that the axis labels and figure titles have been edited and/or rescaled.

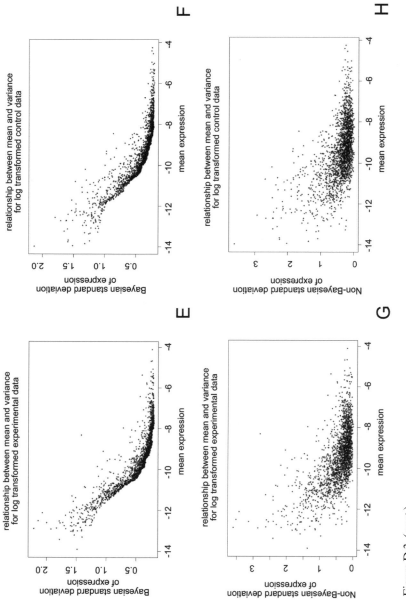

Figure D.3. (cont.)

CyberT generates three output files, two of which (*allgenes.txt* and *siggenes.txt*) can be viewed in the browser window and/or downloaded and imported into a spreadsheet application for user-specific formatting and data manipulation. The *siggenes* file is like the *allgenes* file except it contains data for only those genes that pass the stringent Bonferroni test – if any. These files contain the original data, followed by a number of additional columns containing summary statistics such as: the mean and standard deviation of both raw and log-transformed data, estimates of the standard deviations employing the Bayesian prior, t-tests incorporating the Bayesian prior on both the raw and log-transformed data, p-values associated with t-tests, and "signed fold change" for each gene. In addition to the *allgenes* and *siggenes* files, an additional postscript or pdf file, *CyberT.ps (pdf)*, is generated. *CyberT.ps* contains a self-explanatory set of six graphs useful in visualizing the data submitted to the program. Figure D.3A–F contains an example of these plots. In some cases, two additional graphs are generated (not shown) that plot the p-values of genes which pass the Bonferroni test against their absolute fold change in expression. However, these plots are suppressed if only a few genes pass the Bonferroni test, which will often be the case.

We continue to incorporate improvements into the CyberT program. For example, we are currently adding the capability to compute posterior probabilities of differential expression (PPDE values described in Chapter 7) into this program. When this is completed an additional column with the PPDE value for every gene will appear in the output files.

REFERENCES

1. Long, A. D., Mangalam, H. J., Chan, B. Y. P., Tolleri, L., Hatfield, G. W., and Baldi, P. Improved statistical inference from DNA microarray data using analysis of variance and a Bayesian statistical framework: analysis of global gene expression in *Escherichia coli* K12. 2001. *Journal of Biological Chemistry* 276(23):19937–19944.
2. Baldi, P., and Long, A. D. A Bayesian framework for the analysis of microarray expression data: regularized t-test and statistical inferences of gene changes. 2001. *Bioinformatics* 17(6):509–519.
3. Arfin, S. M., Long, A. D., Ito, E. T., Tolleri, L., Riehle, M. M., Paegle, E. S., and Hatfield, G. W. Global gene expression profiling in *Escherichia coli* K12: the effects of integration host factor. 2000. *Journal of Biological Chemistry* 275(38):29672–29684.

Index

Page numbers in italics indicate a reference to a figure or table.